THE NORTH POLE
AND
INNER EARTH CHRONICLES

★ TWO BOOKS IN ONE - LANDMARK EDITION ★

FEATURING:

THE SMOKY GOD:
A VOYAGE TO THE INNER WORLD

by

WILLIS GEORGE EMERSON

THE SECRET DIARY
OF
ADMIRAL RICHARD E. BYRD

WILD REMNANT PUBLISHING

2023

The North Pole and Inner Earth Chronicles
Two Books in One - Landmark Edition, © Copyright 2023

Foreword by Sebastien Richard

The Smoky God: A Voyage to the Inner World
By Willis George Emerson (and Olaf Jansen), 1908

THIS WORK IS OF THE PUBLIC DOMAIN

The Secret Diary of Admiral Richard E. Byrd
By Admiral Richard E. Byrd, 1947-1956

THIS WORK IS OF THE PUBLIC DOMAIN

Interior formatting by: https://christianbookeditor.uk/

Wild Remnant Publishing
Contact at: wildremnantpublishing@gmail.com

ISBN: 978-1-7777784-9-1

To All the Brave Explorers

Of Our Majestic Realm

&

To Those Who Dare to Venture

Past the Boundaries of Accepted Knowledge

"Great is the Lord, and greatly to be praised in the city of our God, in the mountain of his holiness. Beautiful for situation, the joy of the whole Earth, is mount Zion, on the sides of the north, the city of the great King."

~Psalm 48:1-2

"What people think about you is not supposed to matter much, so long as you yourself know where the truth lies; but I have found out, as have others who move in and out of newspaper headlines, that on occasion it can matter a good deal. For once you enter the world of headlines you learn there is not one truth but two: the one which you know from the facts; and the one which the public, or at any rate a highly imaginative part of the public, acquires by osmosis."

~Admiral Richard E. Byrd

"Most people have made this mistake of thinking Middle-Earth is a particular kind of Earth or is another planet of the science fiction sort but it's just an old fashioned word for this world we live in, as imagined surrounded by the Ocean."

~J. R. R. Tolkien

TABLE OF CONTENTS

PUBLISHER'S FOREWORD

SINCE 2015, an old and somewhat familiar contention has resurfaced and is now making waves worldwide. People are both amused and shocked by the unexpected resurgence of this old discussion, which is even affecting scientific and religious communities. You see, the belief that Earth is a globe was commonly accepted for more than 500 years, until recently. There are now millions of people who question the true shape of our earthly abode and have become, yes, *flat Earth believers*. Indeed, since 2015, this unorthodox belief has spread and spurred countless conferences, debates, podcasts, documentaries, and books; and its spread shows no signs of slowing down.

Regardless of how one may feel about it, this renewed interest, contention, and scrutiny pertaining to our inhabited world has also led to more interest than ever in our cosmology, geography, climate, and geology. So, along with this raging *flat Earth* trend, more and more people have also looked into yet another mysterious and possible aspect of the Earth: its interior. And the question this inevitably leads to is: *Is the Earth hollow?*

And so, within the pages of *The North Pole and Inner Earth Chronicles,* a captivating amalgamation awaits—the fusion of two illustrious narratives that beckon its readers to venture beyond the realms of the familiar and plunge into the uncharted depths of our Earth.

The Smoky God, written by Willis George Emerson back in 1908, weaves a narrative that defies the conventional boundaries of exploration knowledge. Transporting readers to an age where tales of adventure held sway, Emerson's account paints a vivid picture of an unforeseen Father and son voyage across and beneath the Arctic ice. Amidst the mystique lies the narrative attributed to Olaf Jansen, the son, whose deathbed confession purportedly revealed cryptic insights into the realms explored within *The Smoky God.* Within this timeless tale lies the revelation of an unsuspecting journey into the heart of the Earth, divulging a realm teeming with sophistication and an advanced civilisation. Emerson's tale ignites the spark of imagination and curiosity, inviting speculation about the concealed frontiers beneath our very feet.

Aptly complementing this gripping account is the saga of Admiral Richard Evelyn Byrd; his lost diary, whose veiled secrets, whispered through time, add further layers to the anthology. Byrd's enigmatic account, long obscured from public view, remains to this day a subject of fervent speculation and debate. The elusive nature and mysterious contents have sparked intense curiosity, notably heightened by the pivotal events of May 8, 1996, when Byrd's diary surfaced, unravelling an extraordinary revelation on the seventieth anniversary of his historic North Pole flight.

Admiral Richard E. Byrd stands as a prominent and highly respected figure in the chronicles of Arctic and Antarctic exploration. Concealed within the private confines of his *secret diary* lies enigmatic encounters that challenge established perceptions of knowledge. No doubt you will find, as I have, that his account blurs the lines between reality, history, and speculation.

The North Pole and Inner Earth Chronicles therefore stand as a testament to the union of speculative storytelling and documented exploration. This unique compilation serves as a portal to unlock some of the mysteries veiled within one of Earth's innermost sanctums—the North Pole. As you journey through these pages, you're invited to traverse the borders of conventional understanding and embark on a quest into uncharted territories—both geographical and intellectual.

Emerson's evocative narrative and Byrd's enthralling diary converge within this compilation, inviting you to embrace the unknown, question conventional beliefs, and venture into the enigmatic depths where the boundaries of reality and the unexplored interlace.

I, for one, was introduced to *The Secret Diary of Admiral Byrd* back in 2010. A friend who is interested in fringe research, like me, sent me a webpage link that allowed me to read Byrd's remarkable account.

As I delved deeper, it didn't take long for me to find the correlation between Byrd's diary and another captivating and yet controversial book: *The Smoky God: A Voyage to the Inner World*.

When I finally read Willis George Emerson's book years later, I was astounded by the account and its uncanny similarities to Admiral Byrd's secret diary.

Consequently, when I made the decision to publish *The Secret Diary of Admiral Byrd*, I opted to combine it with *The Smoky God*. The two works seamlessly combine to create a captivating resource for speculative and fringe research, appealing to readers, investigators, and the curious.

The Magnetic and Enigmatic North Pole

The unexplored areas of our Earth have always held a fascination in the minds of men. After all, when God commanded man to subdue the Earth and have dominion, back in Genesis 1:26-28; His decree programmed an innate desire for *search and exploration* right into our DNA. Since then, mankind has spared no effort in conquering unexplored territories. And the North Pole, along with the South Pole, is the ultimate unfamiliar destination on Earth. Men who are curious and adventurous are naturally attracted to its unfamiliarity and potential hidden treasures.

Indeed, with its freezing and inhospitable temperatures, infinite ice, endless nights, interminable days, and its northern lights, the North Pole is truly a place of wonder.

Furthermore, the Bible mentions the North as a wondrous place of God's glory, mystery, and rulership:

- Job 26:7: "He stretches out *the north* over empty space and hangs the Earth on nothing."

- Job 37:22: "Out of *the north* comes golden splendor; around God is awesome majesty."
- Isaiah 14:13: "For thou hast said in thine heart, I will ascend into heaven, I will exalt my throne above the stars of God: I will sit also upon the mount of the congregation, *in the sides of the north.*"
- Ezekiel 1:4: "As I looked, behold, a storm wind was *coming from the north,* a great cloud with fire flashing forth continually and a bright light around it, and in its midst something like glowing metal in the midst of the fire." Was Ezekiel's mysterious vision of God preceded by some Northern Lights *"coming from the north?"*
- Psalm 48:1-2: "Great is the Lord, and greatly to be praised in the city of our God, in the mountain of his holiness. Beautiful for situation, the joy of the whole Earth, is mount Zion, *on the sides of the north,* the city of the great King."

Who Are the Mysterious Beings Featured in These Accounts?

Both *The Smoky God* and *The Secret Diary of Admiral Byrd* discuss an advanced civilization accessed through the North Pole and found inside the Earth. Both accounts also unquestionably refer to a mysterious *inner Earth* (aka Hollow Earth). But is there such a thing? Is this even a possibility?

The ancient texts of Scripture again support the believability of this notion. Even in passages deemed familiar, such as the Ten Commandments found in Exodus 20, there are sections that can make one question the boundaries of our reality. For instance, what is meant by this wording in the second commandment?

"Thou shalt not make unto thee any graven image, or any likeness of any thing that is in heaven above, ***or that is in the Earth beneath, or that is in the water under the Earth."*** ~Exodus 20:4

We know that the creation of graven images by any society usually stems from a desire to worship a higher power. It appears that when giving commandments to Jacob/Israel, Yahweh took into account the presence of creatures living beneath the Earth. At the very least, He knew that certain beings beneath the surface could spark our fascination, perhaps even our admiration and worship. I find this notion both fascinating and troubling.

So, who could the civilization spoken of in *The Secret Diary of Admiral Byrd* and Olaf Jansen in *The Smoky God* be? Who did these two men meet past the frozen North Pole? Let us consider some of the possibilities, shall we?

1. **Ancient Aliens?** Admiral Byrd and Olaf Jansen's accounts are frequently used by UFO researchers as proof of an extraterrestrial civilization observing human activity on Earth. Certainly, the main fallacy in this theory is that they wouldn't be from outer space, but rather from *within* the Earth. This concept would classify these beings as intra-terrestrials, which is quite intriguing.

2. **A More Advanced Humanoid Species?** Another theory suggests that these two explorers encountered a

previously unknown humanoid species. Some proponents of this theory have advanced that this civilization might be the relocated people of Atlantis, who built their inner Earth sanctum after the lost continent's demise. While this thesis holds water, pardon the pun, it is still only conjecture and remains puzzling, especially considering the sheer size of the beings, as recounted by Olaf Jansen. Furthermore, in Admiral Byrd's account, the place is called *Arianni* by 'the Master', and not Atlantis. This, coupled with the Swastikas on their flying crafts, as noted by Admiral Byrd, points to some type of mysterious Arian race of super-advanced beings. Regardless, the supporters of this thesis would also need to provide further elucidation of how the inhabitants, if indeed Atlantean, survived the destruction of their lost city-continent of Atlantis.

3. **Demonic Deceivers or Nephilim?** Some have theorized that both Olaf Jansen and Admiral Richard E. Byrd encountered supernatural beings—possibly demons, fallen angels, or the remnant of the Nephilim spoken of in Genesis 6 and *The Book of Enoch*. While this is an interesting possibility, one would have to wonder why both men remained unharmed and were even allowed to leave their abode and return to the surface world. It could be argued, of course, that this was done to enable the spread of falsehoods to us surface dwellers who read their accounts.

4. **The Throne of God, Jesus, and the Camp of the Saints?** There is another theory, more recent, which suggests that what these explorers encountered could be the lost

garden of Eden, the camp of the Saints, Mount Zion's interior, or even the New Jerusalem. Indeed, Olaf Jansen, in *The Smoky God,* referred to the place where he sojourned as "Eden". While I am fascinated by this concept, it prompts theological uncertainties. At the forefront is the question of why Jesus Christ and/or the saints would choose to remain hidden in "Eden" while we suffer under wicked elitist rulers on the Earth (?). As an answer, the proponents of this particular theory have suggested that we are probably living in *Satan's little season* (a short season of worldwide Satanic deception ordained by God) as expressed in Revelation chapter 20 verses 3 & 7-9. This, of course, is just the gist of it—there are many more details to this theory that I will not address here. And while undeniably intriguing, this assertion raises even more head-scratching questions. To begin with, one of those is, once again, how can we account for the massive size of the people encountered in *The Smoky God?* This issue is puzzling, whether we assume this is how resurrected bodies look or not. The theory becomes more questionable as neither account includes beings who identify as any of the ancient saints or Christ Himself, leaving their true identity as a perplexing question.

Without a doubt, the aforementioned theories all offer valid points worth considering. And we could probably go on and add some more to the lot. It is then safe to say that your guess and preference are as good as mine. As we progress in uncovering more truth, however, we will undoubtedly find additional answers; hopefully in the months and years ahead.

In the meantime, as you immerse yourself in *The North Pole and Inner Earth Chronicles,* a convergence of history, speculation, and enigma awaits—uniting the realms of the known and the mysterious.

I hope these accounts entertain you, for sure. But primarily, I want them to stimulate your imagination, provoke questions, and motivate you to find answers. Given the rampant deception these days, what we really need are answers, and hopefully, *more truth.* In the end, the truth, surely, shall set us free. Perhaps it is only when that time comes that we will fully grasp what Admiral Byrd tried to convey when he wrote:

"Just as the long night of the Arctic ends, the brilliant sunshine of Truth shall come again... and those who are of darkness shall fall in its Light... for I have seen that land beyond the Pole, that Center of the Great Unknown."

~Admiral Richard E. Byrd

Sebastien Richard

THE SMOKY GOD: A VOYAGE TO THE INNER WORLD

"He is the God who sits in the center, on the navel of the Earth, and He is the interpreter of religion to all mankind." ~Plato

PART ONE: AUTHOR'S FOREWORD

I FEAR the seemingly incredible story which I am about to relate will be regarded as the result of a distorted intellect superinduced, possibly, by the glamour of unveiling a marvelous mystery, rather than a truthful record of the unparalleled experiences related by one Olaf Jansen, whose eloquent madness so appealed to my imagination that all thought of an analytical criticism has been effectually dispelled.

Marco Polo will doubtless shift uneasily in his grave at the strange story I am called upon to chronicle; a story as strange as a Munchausen tale. It is also incongruous that I, a disbeliever, should be the one to edit the story of Olaf Jansen, whose name is now for the first time given to the world, yet who must hereafter rank as one of the notables of Earth.

I freely confess his statements admit of no rational analysis, but have to do with the profound mystery concerning the frozen North that for centuries has claimed the attention of scientists and laymen alike.

However much they are at variance with the cosmographical manuscripts of the past, these plain statements may be relied upon as a record of the things Olaf Jansen claims to have seen with his own eyes.

A hundred times I have asked myself whether it is possible that the world's geography is incomplete, and that the startling narrative of Olaf Jansen is predicated upon demonstrable facts. The reader may be able to answer these queries to his own satisfaction, however far the chronicler of this narrative may be from having reached a conviction. Yet sometimes even I am at a loss to know whether I have been led away from an abstract truth by the ignes fatui of a clever superstition, or whether heretofore accepted facts are, after all, founded upon falsity.

It may be that the true home of Apollo was not at Delphi, but in that older Earth-center of which Plato speaks, where he says: "Apollo's real home is among the Hyperboreans, in a land of perpetual life, where mythology tells us two doves flying from the two opposite ends of the world met in this fair region, the home of Apollo. Indeed, according to Hecataeus, Leto, the mother of Apollo, was born on an island in the Arctic Ocean far beyond the North Wind."

It is not my intention to attempt a discussion of the theogony of the deities nor the cosmogony of the world. My simple duty is to enlighten the world concerning a heretofore unknown portion of the universe, as it was seen and described by the old Norseman, Olaf Jansen.

Interest in northern research is international. Eleven nations are engaged in, or have contributed to, the perilous work of trying to solve Earth's one remaining cosmological mystery.

There is a saying, ancient as the hills, that *"truth is stranger than fiction,"* and in a most startling manner has this axiom been brought home to me within the last fortnight.

It was just two o'clock in the morning when I was aroused from a restful sleep by the vigorous ringing of my door-bell. The untimely disturber proved to be a messenger bearing a note, scrawled almost to the point of illegibility, from an old Norseman by the name of Olaf Jansen. After much deciphering, I made out the writing, which simply said: *"Am ill unto death. Come."* The call was imperative, and I lost no time in making ready to comply.

Perhaps I may as well explain here that Olaf Jansen, a man who quite recently celebrated his ninety-fifth birthday, has for the last half-dozen years been living alone in an unpretentious bungalow out Glendale way, a short distance from the business district of Los Angeles, California.

It was less than two years ago, while out walking one afternoon that I was attracted by Olaf Jansen's house and its homelike surroundings, toward its owner and occupant, whom I afterward came to know as a believer in the ancient worship of Odin and Thor.

There was a gentleness in his face, and a kindly expression in the keenly alert gray eyes of this man who had lived more than four-score years and ten; and, withal, a sense of loneliness that appealed to my sympathy. Slightly stooped, and with his hands clasped behind him, he walked back and forth with slow and measured tread, that day when first we met. I can hardly say what particular motive impelled me to pause in my walk and engage him in conversation. He seemed pleased when

I complimented him on the attractiveness of his bungalow, and on the well-tended vines and flowers clustering in profusion over its windows, roof and wide piazza.

I soon discovered that my new acquaintance was no ordinary person, but one profound and learned to a remarkable degree; a man who, in the later years of his long life, had dug deeply into books and become strong in the power of meditative silence.

I encouraged him to talk, and soon gathered that he had resided only six or seven years in Southern California, but had passed the dozen years prior in one of the middle Eastern states. Before that he had been a fisherman off the coast of Norway, in the region of the Lofoden Islands, from whence he had made trips still farther north to Spitzbergen and even to Franz Josef Land.

When I started to take my leave, he seemed reluctant to have me go, and asked me to come again. Although at the time I thought nothing of it, I remember now that he made a peculiar remark as I extended my hand in leave-taking. "You will come again?" he asked. "Yes, you will come again some day. I am sure you will; and I shall show you my library and tell you many things of which you have never dreamed, things so wonderful that it may be you will not believe me."

I laughingly assured him that I would not only come again, but would be ready to believe whatever he might choose to tell me of his travels and adventures.

In the days that followed I became well acquainted with Olaf Jansen, and, little by little, he told me his story, so marvelous, that its very daring challenges reason and belief. The old Norseman always expressed himself with so much earnestness and sincerity that I became enthralled by his strange narrations.

Then came the messenger's call that night, and within the hour I was at Olaf Jansen's bungalow.

He was very impatient at the long wait, although after being summoned I had come immediately to his bedside.

"I must hasten," he exclaimed, while yet he held my hand in greeting. "I have much to tell you that you know not, and I will trust no one but you. I fully realize," he went on hurriedly, "that I shall not survive the night. The time has come to join my fathers in the great sleep."

I adjusted the pillows to make him more comfortable, and assured him I was glad to be able to serve him in any way possible, for I was beginning to realize the seriousness of his condition.

The lateness of the hour, the stillness of the surroundings, the uncanny feeling of being alone with the dying man, together with his weird story, all combined to make my heart beat fast and loud with a feeling for which I have no name. Indeed, there were many times that night by the old Norseman's couch, and there have been many times since, when a sensation rather than a conviction took possession of my very soul, and I seemed not only to believe in, but actually see, the strange lands, the strange people and the strange world of which he told, and to hear the mighty orchestral chorus of a thousand lusty voices.

For over two hours he seemed endowed with almost superhuman strength, talking rapidly, and to all appearances, rationally. Finally he gave into my hands certain data, drawings and crude maps. "These," said he in conclusion, "I leave in your hands. If I can have your promise to give them to the world, I shall die happy, because I desire that people may know the truth, for then all mystery concerning the frozen Northland will be explained. There is no chance of your suffering the fate I suffered. They will not put you in irons, nor confine you in a mad-house, because you are not telling your own story, but mine, and I, thanks to the gods, Odin and Thor, will be in my grave, and so beyond the reach of disbelievers who would persecute."

Without a thought of the farreaching results the promise entailed, or foreseeing the many sleepless nights which the obligation has since brought me, I gave my hand and with it a pledge to discharge faithfully his dying wish.

As the sun rose over the peaks of the San Jacinto, far to the eastward, the spirit of Olaf Jansen, the navigator, the explorer and worshiper of Odin and Thor, the man whose experiences and travels, as related, are without a parallel in all the world's history, passed away, and I was left alone with the dead.

And now, after having paid the last sad rites to this strange man from the Lofoden Islands, and the still farther "Northward Ho!", the courageous explorer of frozen regions, who in his declining years (after he had passed the four-score mark) had sought an asylum of restful peace in sun-favored California, I will undertake to make public his story.

But, first of all, let me indulge in one or two reflections:

Generation follows generation, and the traditions from the misty past are handed down from sire to son, but for some strange reason interest in the ice-locked unknown does not abate with the receding years, either in the minds of the ignorant or the tutored.

With each new generation a restless impulse stirs the hearts of men to capture the veiled citadel of the Arctic, the circle of silence, the land of glaciers, cold wastes of waters and winds that are strangely warm. Increasing interest is manifested in the mountainous icebergs, and marvelous speculations are indulged in concerning the Earth's center of gravity, the cradle of the tides, where the whales have their nurseries, where the magnetic needle goes mad, where the Aurora Borealis illumines the night, and where brave and courageous spirits of every generation dare to venture and explore, defying the dangers of the "Farthest North."

One of the ablest works of recent years is "Paradise Found, or the Cradle of The Human Race at the North Pole," by William F. Warren. In his carefully prepared volume, Mr. Warren almost stubbed his toe against the real truth, but missed it seemingly by only a hair's breadth, if the old Norseman's revelation be true.

Dr. Orville Livingston Leech, scientist, in a recent article, says:

"The possibilities of a land inside the Earth were first brought to my attention when I picked up a geode on the shores of the Great Lakes. The geode is a spherical and apparently solid stone, but when broken is found to be hollow and coated with crystals. The Earth is only

a larger form of a geode, and the law that created the geode in its hollow form undoubtedly fashioned the Earth in the same way."

In presenting the theme of this almost incredible story, as told by Olaf Jansen, and supplemented by manuscript, maps and crude drawings entrusted to me, a fitting introduction is found in the following quotation:

"In the beginning God created the heaven and the Earth, and the Earth was without form and void." And also, *"God created man in his own image."* Therefore, even in things material, man must be God-like, because he is created in the likeness of the Father.

A man builds a house for himself and family. The porches or verandas are all without, and are secondary. The building is really constructed for the conveniences within.

Olaf Jansen makes the startling announcement through me, an humble instrument, that in like manner, God created the Earth for the "within"—that is to say, for its lands, seas, rivers, mountains, forests and valleys, and for its other internal conveniences, while the outside surface of the Earth is merely the veranda, the porch, where things grow by comparison but sparsely, like the lichen on the mountain side, clinging determinedly for bare existence.

Take an egg-shell, and from each end break out a piece as large as the end of this pencil. Extract its contents, and then you will have a perfect representation of Olaf Jansen's Earth. The distance from the inside surface to the outside surface, according to him, is about three hundred miles. The center of gravity is not in the center of the Earth, but in the center of the shell or crust; therefore, if the thickness of the Earth's crust or shell is three hundred miles, the center of gravity is one hundred and fifty miles below the surface.

In their log-books Arctic explorers tell us of the dipping of the needle as the vessel sails in regions of the farthest north known. In reality, they are at the curve; on the edge of the shell, where gravity is geometrically increased, and while the electric current seemingly dashes off into space toward the phantom idea of the North Pole, yet this same electric current drops again and continues its course southward along the inside surface of the Earth's crust.

In the appendix to his work, Captain Sabine gives an account of experiments to determine the acceleration of the pendulum in different latitudes. This appears to have resulted from the joint labor of Peary and Sabine. He says: "The accidental discovery that a pendulum on being removed from Paris to the neighborhood of the equator increased its time of vibration, gave the first step to our present knowledge that the polar axis of the globe is less than the equatorial; that the force of gravity at the surface of the Earth increases progressively from the equator toward the poles."

According to Olaf Jansen, in the beginning this old world of ours was created solely for the "within" world, where are located the four great rivers—the Euphrates, the Pison, the Gihon and the Hiddekel. These same names of rivers, when applied to streams on the "outside" surface of the Earth, are purely traditional from an antiquity beyond the memory of man.

On the top of a high mountain, near the fountain-head of these four rivers, Olaf Jansen, the Norseman, claims to have discovered the long-lost "Garden of Eden," the veritable navel of the Earth, and to have spent over two years studying and reconnoitering in this marvelous "within" land, exuberant with stupendous plant life and abounding in giant animals; a land where the people live to be centuries old, after the order of Methuselah and other Biblical characters; a region where one-quarter of the "inner" surface is water and three-quarters land; where there are large oceans and many rivers and lakes; where the cities are superlative in construction and magnificence; where modes of transportation are as far in advance of ours as we with our boasted achievements are in advance of the inhabitants of "darkest Africa."

The distance directly across the space from inner surface to inner surface is about six hundred miles less than the recognized diameter of the Earth. In the identical center of this vast vacuum is the seat of electricity—a mammoth ball of dull red fire—not startlingly brilliant, but surrounded by a white, mild, luminous cloud, giving out uniform warmth, and held in its place in the center of this internal space by the immutable law of gravitation. This electrical cloud is known to the people "within" as the abode of "The Smoky God." They believe it to be the throne of "The Most High."

Olaf Jansen reminded me of how, in the old college days, we were all familiar with the laboratory demonstrations of centrifugal motion, which clearly proved that, if the Earth were a solid, the rapidity of its revolution upon its axis would tear it into a thousand fragments.

The old Norseman also maintained that from the farthest points of land on the islands of Spitzbergen and Franz Josef Land, flocks of geese may be seen annually flying still farther northward, just as the sailors and explorers record in their log-books. No scientist has yet been audacious enough to attempt to explain, even to his own satisfaction, toward what lands these winged fowls are guided by their subtle instinct. However, Olaf Jansen has given us a most reasonable explanation.

The presence of the open sea in the Northland is also explained. Olaf Jansen claims that the northern aperture, intake or hole, so to speak, is about fourteen hundred miles across. In connection with this, let us read what Explorer Nansen writes, on page 288 of his book: "I have never had such a splendid sail. On to the north, steadily north, with a good wind, as fast as steam and sail can take us, an open sea mile after mile, watch after watch, through these unknown regions, always clearer and clearer of ice, one might almost say: 'How long will it last?' The eye always turns to the northward as one paces the bridge. It is gazing into the future. But there is always the same dark sky ahead which means open sea." Again, the Norwood Review of England, in its issue of May 10, 1884, says: "We do not admit that there is ice up to the Pole—once inside the great ice barrier, a new world breaks upon the explorer, the climate is mild like that of England, and, afterward, balmy as the Greek Isles."

Some of the rivers "within," Olaf Jansen claims, are larger than our Mississippi and Amazon rivers combined, in point of volume of water carried; indeed their greatness is occasioned by their width and depth rather than their length, and it is at the mouths of these mighty rivers, as they flow northward and southward along the inside surface of the Earth, that mammoth icebergs are found, some of them fifteen and twenty miles wide and from forty to one hundred miles in length.

Is it not strange that there has never been an iceberg encountered either in the Arctic or Antarctic Ocean that is not composed of fresh water? Modern scientists claim that freezing eliminates the salt, but Olaf Jansen claims differently.

Ancient Hindoo, Japanese and Chinese writings, as well as the hieroglyphics of the extinct races of the North American continent, all speak of the custom of sun-worshiping, and it is possible, in the startling light of Olaf Jansen's revelations, that the people of the inner world, lured away by glimpses of the sun as it shone upon the inner surface of the Earth, either from the northern or the southern opening, became dissatisfied with "The Smoky God," the great pillar or mother cloud of electricity, and, weary of their continuously mild and pleasant atmosphere, followed the brighter light, and were finally led beyond the ice belt and scattered over the "outer" surface of the Earth, through Asia, Europe, North America and, later, Africa, Australia and South America. (1)

(1 The following quotation is significant; "It follows that man issuing from a mother-region still undetermined but which a number of considerations indicate to have been in the North, has radiated in several directions; that his migrations have been constantly from North to South."—M. le Marquis G. de Saporta, in Popular Science Monthly, October, 1883, page 753.)

It is a notable fact that, as we approach the Equator, the stature of the human race grows less. But the Patagonians of South America are probably the only aborigines from the center of the Earth who came out through the aperture usually designated as the South Pole, and they are called the giant race.

Olaf Jansen avers that, in the beginning, the world was created by the Great Architect of the Universe, so that man might dwell upon its "inside" surface, which has ever since been the habitation of the "chosen."

They who were driven out of the "Garden of Eden" brought their traditional history with them.

The history of the people living "within" contains a narrative suggesting the story of Noah and the ark with which we are familiar. He sailed away, as did Columbus, from a certain port, to a strange land he had heard of far to the northward, carrying with him all manner of beasts of the fields and fowls of the air, but was never heard of afterward.

On the northern boundaries of Alaska, and still more frequently on the Siberian coast, are found boneyards containing tusks of ivory in quantities so great as to suggest the burying-places of antiquity. From Olaf Jansen's account, they have come from the great prolific animal life that abounds in the fields and forests and on the banks of numerous rivers of the Inner World. The materials were caught in the ocean currents, or were carried on ice-floes, and have accumulated like driftwood on the Siberian coast. This has been going on for ages, and hence these mysterious bone-yards.

On this subject William F. Warren, in his book already cited, pages 297 and 298, says: "The Arctic rocks tell of a lost Atlantis more wonderful than Plato's. The fossil ivory beds of Siberia excel everything of the kind in the world. From the days of Pliny, at least, they have constantly been undergoing exploitation, and still they are the chief headquarters of supply. The remains of mammoths are so abundant that, as Gratacap says, 'the northern islands of Siberia seem built up of crowded bones.' Another scientific writer, speaking of the islands of New Siberia, northward of the mouth of the River Lena, uses this language: 'Large quantities of ivory are dug out of the ground every year. Indeed, some of the islands are believed to be nothing but an accumulation of drift-timber and the bodies of mammoths and other antediluvian animals frozen together.' From this we may infer that, during the years that have elapsed since the Russian conquest of Siberia, useful tusks from more than twenty thousand mammoths have been collected."

But now for the story of Olaf Jansen. I give it in detail, as set down by himself in manuscript, and woven into the tale, just as he placed them, are certain quotations from recent works on Arctic exploration, showing how carefully the old Norseman compared with his own experiences those of other voyagers to the frozen North. Thus wrote the disciple of Odin and Thor...

PART TWO: OLAF JANSEN'S STORY

MY name is Olaf Jansen. I am a Norwegian, although I was born in the little seafaring Russian town of Uleaborg, on the eastern coast of the Gulf of Bothnia, the northern arm of the Baltic Sea.

My parents were on a fishing cruise in the Gulf of Bothnia, and put into this Russian town of Uleaborg at the time of my birth, being the twenty-seventh day of October, 1811.

My father, Jens Jansen, was born at Rodwig on the Scandinavian coast, near the Lofoden Islands, but after marrying made his home at Stockholm, because my mother's people resided in that city. When seven years old, I began going with my father on his fishing trips along the Scandinavian coast.

Early in life I displayed an aptitude for books, and at the age of nine years was placed in a private school in Stockholm, remaining there until I was fourteen. After this I made regular trips with my father on all his fishing voyages.

My father was a man fully six feet three in height, and weighed over fifteen stone, a typical Norseman of the most rugged sort, and capable of more endurance than any other man I have ever known. He possessed the gentleness of a woman in tender little ways, yet his determination and will-power were beyond description. His will admitted of no defeat.

I was in my nineteenth year when we started on what proved to be our last trip as fishermen, and which resulted in the strange story that shall be given to the world,—but not until I have finished my Earthly pilgrimage.

I dare not allow the facts as I know them to be published while I am living, for fear of further humiliation, confinement and suffering. First of all, I was put in irons by the captain of the whaling vessel that rescued me, for no other reason than that I told the truth about the marvelous discoveries made by my father and myself. But this was far from being the end of my tortures.

After four years and eight months' absence I reached Stockholm, only to find my mother had died the previous year, and the property left by my parents in the possession of my mother's people, but it was at once made over to me.

All might have been well, had I erased from my memory the story of our adventure and of my father's terrible death.

Finally, one day I told the story in detail to my uncle, Gustaf Osterlind, a man of considerable property, and urged him to fit out an expedition for me to make another voyage to the strange land.

At first I thought he favored my project. He seemed interested, and invited me to go before certain officials and explain to them, as I had to him, the story of our travels and discoveries. Imagine my disappointment and horror when, upon the conclusion of my narrative, certain papers were signed by my uncle, and, without warning, I found myself arrested and hurried away to dismal and fearful confinement in a madhouse, where I remained for twenty-eight years—long, tedious, frightful years of suffering!

I never ceased to assert my sanity, and to protest against the injustice of my confinement. Finally, on the seventeenth of October, 1862, I was released. My uncle was dead, and the friends of my youth were now strangers. Indeed, a man over fifty years old, whose only known record is that of a madman, has no friends.

I was at a loss to know what to do for a living, but instinctively turned toward the harbor where fishing boats in great numbers were anchored, and within a week I had shipped with a fisherman by the name of Yan Hansen, who was starting on a long fishing cruise to the Lofoden Islands.

Here my earlier years of training proved of the very greatest advantage, especially in enabling me to make myself useful. This was but the beginning of other trips, and by frugal economy I was, in a few years, able to own a fishing-brig of my own. For twenty-seven years thereafter I followed the sea as a fisherman, five years working for others, and the last twenty-two for myself.

During all these years I was a most diligent student of books, as well as a hard worker at my business, but I took great care not to mention to anyone the story concerning the discoveries made by my father and myself. Even at this late day I would be fearful of having any one see or know the things I am writing, and the records and maps I have in my keeping.

When my days on Earth are finished, I shall leave maps and records that will enlighten and, I hope, benefit mankind.

The memory of my long confinement with maniacs, and all the horrible anguish and sufferings are too vivid to warrant my taking further chances.

In 1889 I sold out my fishing boats, and found I had accumulated a fortune quite sufficient to keep me the remainder of my life. I then came to America.

For a dozen years my home was in Illinois, near Batavia, where I gathered most of the books in my present library, though I brought many choice volumes from Stockholm. Later, I came to Los Angeles, arriving here March 4, 1901. The date I well remember, as it was President McKinley's second inauguration day. I bought this humble home and determined, here in the privacy of my own abode, sheltered by my own vine and fig-tree, and with my books about me, to make maps and drawings of the new lands we had discovered, and also to write the story in detail from the time my father and I left Stockholm until the tragic event that parted us in the Antarctic Ocean.

I well remember that we left Stockholm in our fishing-sloop on the third day of April, 1829, and sailed to the southward, leaving Gothland Island to the left and Oeland Island to the right. A few days later we succeeded in doubling Sandhommar Point, and made our way through the sound which separates Denmark from the Scandinavian coast. In due time we put in at the town of Christiansand, where we rested two days, and then started around the Scandinavian coast to the westward, bound for the Lofoden Islands.

My father was in high spirit, because of the excellent and gratifying returns he had received from our last catch by marketing at Stockholm, instead of selling at one of the seafaring towns along the Scandinavian coast. He was especially pleased with the sale of some ivory tusks that he had found on the west coast of Franz Joseph Land during one of his northern cruises the previous year, and he expressed the hope that this time we might again be fortunate enough to load our little fishing-sloop with ivory, instead of cod, herring, mackerel and salmon.

We put in at Hammerfest, latitude seventy-one degrees and forty minutes, for a few days' rest. Here we remained one week, laying in an extra supply of provisions and several casks of drinking-water, and then sailed toward Spitzbergen.

For the first few days we had an open sea and a favoring wind, and then we encountered much ice and many icebergs. A vessel larger than our little fishing-sloop could not possibly have threaded its way among the labyrinth of icebergs or squeezed through the barely open channels. These monster bergs presented an endless succession of crystal palaces, of massive cathedrals and fantastic mountain ranges, grim and sentinel-like, immovable as some towering cliff of solid rock, standing; silent as a sphinx, resisting the restless waves of a fretful sea.

After many narrow escapes, we arrived at Spitzbergen on the 23d of June, and anchored at Wijade Bay for a short time, where we were quite successful in our catches. We then lifted anchor and sailed through the Hinlopen Strait, and coasted along the North-East-Land. (2)

(2 It will be remembered that Andree started on his fatal balloon voyage from the northwest coast of Spitzbergen.)

A strong wind came up from the southwest, and my father said that we had better take advantage of it and try to reach Franz Josef Land, where, the year before he had, by accident, found the ivory tusks that had brought him such a good price at Stockholm.

Never, before or since, have I seen so many sea-fowl; they were so numerous that they hid the rocks on the coast line and darkened the sky.

For several days we sailed along the rocky coast of Franz Josef Land. Finally, a favoring wind came up that enabled us to make the West Coast, and, after sailing twenty-four hours, we came to a beautiful inlet.

One could hardly believe it was the far Northland. The place was green with growing vegetation, and while the area did not comprise more than one or two acres, yet the air was warm and tranquil. It seemed to be at that point where the Gulf Stream's influence is most keenly felt. (3)

(3 Sir John Barrow, Bart., F.R.S., in his work entitled "Voyages of Discovery and Research Within the Arctic Regions," says on page 57: "Mr. Beechey refers to what has frequently been found and noticed—the mildness of the temperature on the western coast of Spitzbergen, there being little or no sensation of cold, though the thermometer might be only a few degrees above the freezing-point. The brilliant and lively effect of a clear day, when the sun shines forth with a pure sky, whose azure hue is so intense as to find no parallel even in the boasted Italian sky.")

On the east coast there were numerous icebergs, yet here we were in open water. Far to the west of us, however, were icepacks, and still farther to the westward the ice appeared like ranges of low hills. In front of us, and directly to the north, lay an open sea. (4)

(4 Captain Kane, on page 299, quoting from Morton's Journal on Monday, the 26th of December, says: "As far as I could see, the open passages were fifteen miles or more wide, with sometimes mashed ice separating them. But it is all small ice, and I think it either drives out to the open space to the north or rots and sinks, as I could see none ahead to the north.")

My father was an ardent believer in Odin and Thor, and had frequently told me they were gods who came from far beyond the "North Wind."

There was a tradition, my father explained, that still farther northward was a land more beautiful than any that mortal man had ever known, and that it was inhabited by the "Chosen."(5)

(5 We find the following in "Deutsche Mythologie," page 778, from the pen of Jakob Grimm; "Then, the sons of Bor built in the middle of the universe the city called Asgard, where dwell the gods and their kindred, and from that abode work out so many wondrous things both on the Earth and in the heavens above it. There is in that city a place called Illidskjalf, and when Odin is seated there upon his lofty throne he sees over the whole world and discerns all the actions of men.")

My youthful imagination was fired by the ardor, zeal and religious fervor of my good father, and I exclaimed: "Why not sail to this goodly land? The sky is fair, the wind favorable and the sea open."

Even now I can see the expression of pleasurable surprise on his countenance as he turned toward me and asked: "My son, are you willing to go with me and explore—to go far beyond where man has ever ventured?" I answered affirmatively. "Very well," he replied. "May the god Odin protect us!" and, quickly adjusting the sails, he glanced at our compass, turned the prow in due northerly direction through an open channel, and our voyage had begun. (6)

(6 Hall writes, on page 288: "On the 23rd of January the two Esquimaux, accompanied by two of the seamen, went to Cape Lupton. They reported a sea of open water extending as far as the eye could reach.")

The sun was low in the horizon, as it was still the early summer. Indeed, we had almost four months of day ahead of us before the frozen night could come on again.

Our little fishing-sloop sprang forward as if eager as ourselves for adventure. Within thirty-six hours we were out of sight of the highest point on the coast line of Franz Josef Land. We seemed to be in a strong current running north by northeast. Far to the right and to the left of us were icebergs, but our little sloop bore down on the narrows and passed through channels and out into open seas—channels so narrow in places that, had our craft been other than small, we never could have gotten through.

On the third day we came to an island. Its shores were washed by an open sea. My father determined to land and explore for a day. This new land was destitute of timber, but we found a large accumulation of drift-wood on the northern shore.

Some of the trunks of the trees were forty feet long and two feet in diameter. (7)

(7 Greely tells us in vol. 1, page 100, that: "Privates Connell and Frederick found a large coniferous tree on the beach, just above the extreme high-water mark. It was nearly thirty inches in circumference, some thirty feet long, and had apparently been carried to that point by a current within a couple of years. A portion of it was cut up for fire-wood, and for the first time in that valley, a bright, cheery camp-fire gave comfort to man.")

After one day's exploration of the coast line of this island, we lifted anchor and turned our prow to the north in an open sea. (8)

(8 Dr. Kane says, on page 379 of his works: "I cannot imagine what becomes of the ice. A strong current sets in constantly to the north; but, from altitudes of more than five hundred feet, I saw only narrow strips of ice, with great spaces of open water, from ten to fifteen miles in breadth, between them. It must, therefore, either go to an open space in the north, or dissolve.")

I remember that neither my father nor myself had tasted food for almost thirty hours. Perhaps this was because of the tension of excitement about our strange voyage in waters farther north, my father said, than anyone had ever before been. Active mentality had dulled the demands of the physical needs.

Instead of the cold being intense as we had anticipated, it was really warmer and more pleasant than it had been while in Hammerfest on the north coast of Norway, some six weeks before. (9)

(9 Captain Peary's second voyage relates another circumstance which may serve to confirm a conjecture which has long been maintained by some, that an open sea, free of ice, exists at or near the Pole. "On the second of November," says Peary, "the wind freshened up to a gale from north by west, lowered the thermometer before midnight to 5 degrees, whereas, a rise of wind at Melville Island was generally accompanied by a simultaneous rise in the thermometer at low temperatures. May not this," he asks, "be occasioned by the wind blowing over an open sea in the quarter from which the wind blows? And tend to confirm the opinion that at or near the Pole an open sea exists?")

We both frankly admitted that we were very hungry, and forthwith I prepared a substantial meal from our well-stored larder. When we had partaken heartily of the repast, I told my father I believed I would sleep, as I was beginning to feel quite drowsy. "Very well," he replied, "I will keep the watch."

I have no way to determine how long I slept; I only know that I was rudely awakened by a terrible commotion of the sloop. To my surprise, I found my father sleeping soundly. I cried out lustily to him, and starting up, he sprang quickly to his feet. Indeed, had he not instantly clutched the rail, he would certainly have been thrown into the seething waves.

A fierce snow-storm was raging. The wind was directly astern, driving our sloop at a terrific speed, and was threatening every moment to capsize us. There was no time to lose, the sails had to be lowered immediately. Our boat was writhing in convulsions. A few icebergs we knew were on either side of us, but fortunately the channel was open directly to the north. But would it remain so? In front of us, girding the horizon from left to right, was a vaporish fog or mist, black as Egyptian night at the water's edge, and white like a steam-cloud toward the top, which was finally lost to view as it blended with the great white flakes of falling snow. Whether it covered a treacherous iceberg, or some other hidden obstacle against which our little sloop would dash and send us to a watery grave, or was merely the phenomenon of an Arctic fog, there was no way to determine. (10)

(10 On page 284 of his works, Hall writes: "From the top of Providence Berg, a dark fog was seen to the north, indicating water. At 10 a. m. three of the men (Kruger, Nindemann and Hobby) went to Cape Lupton to ascertain if possible the extent of the open water. On their return they reported several open spaces and much young ice—not more than a day old, so thin that it was easily broken by throwing pieces of ice upon it.")

By what miracle we escaped being dashed to utter destruction, I do not know. I remember our little craft creaked and groaned, as if its joints were breaking. It rocked and staggered to and fro as if clutched by some fierce undertow of whirlpool or maelstrom.

Fortunately our compass had been fastened with long screws to a crossbeam. Most of our provisions, however, were tumbled out and swept away from the deck of the cuddy, and had we not taken the precaution at the very beginning to tie ourselves firmly to the masts of the sloop, we should have been swept into the lashing sea.

Above the deafening tumult of the raging waves, I heard my father's voice. "Be courageous, my son," he shouted, "Odin is the god of the waters, the companion of the brave, and he is with us. Fear not."

To me it seemed there was no possibility of our escaping a horrible death. The little sloop was shipping water, the snow was falling so fast as to be blinding, and the waves were tumbling over our counters in reckless white-sprayed fury. There was no telling what instant we should be dashed against some drifting ice-pack. The tremendous swells would heave us up to the very peaks of mountainous waves, then plunge us down into the depths of the sea's trough as if our fishing-sloop were a fragile shell. Gigantic white-capped waves, like veritable walls, fenced us in, fore and aft.

This terrible nerve-racking ordeal, with its nameless horrors of suspense and agony of fear indescribable, continued for more than three hours, and all the time we were being driven forward at fierce speed. Then suddenly, as if growing weary of its frantic exertions, the wind began to lessen its fury and by degrees to die down.

At last we were in a perfect calm. The fog mist had also disappeared, and before us lay an iceless channel perhaps ten or fifteen miles wide, with a few icebergs far away to our right, and an intermittent archipelago of smaller ones to the left.

I watched my father closely, determined to remain silent until he spoke. Presently he untied the rope from his waist and, without saying a word, began working the pumps, which fortunately were not damaged, relieving the sloop of the water it had shipped in the madness of the storm.

He put up the sloop's sails as calmly as if casting a fishing-net, and then remarked that we were ready for a favoring wind when it came. His courage and persistence were truly remarkable.

On investigation we found less than one-third of our provisions remaining, while to our utter dismay, we discovered that our water-casks had been swept overboard during the violent plungings of our boat.

Two of our water-casks were in the main hold, but both were empty. We had a fair supply of food, but no fresh water. I realized at once the awfulness of our position. Presently I was seized with a consuming thirst. "It is indeed bad," remarked my father. "However, let us dry our bedraggled clothing, for we are soaked to the skin. Trust to the god Odin, my son. Do not give up hope."

The sun was beating down slantingly, as if we were in a southern latitude, instead of in the far Northland. It was swinging around, its orbit ever visible and rising higher and higher each day, frequently mist-covered, yet always peering through the lacework of clouds like some fretful eye of fate, guarding the mysterious Northland and jealously watching the pranks of man. Far to our right the rays decking the prisms of icebergs were gorgeous. Their reflections emitted flashes of garnet, of diamond, of sapphire. A pyrotechnic panorama of countless colors and shapes, while below could be seen the green-tinted sea, and above, the purple sky.

PART THREE: BEYOND THE NORTH WIND

I TRIED to forget my thirst by busying myself with bringing up some food and an empty vessel from the hold. Reaching over the side-rail, I filled the vessel with water for the purpose of laving my hands and face. To my astonishment, when the water came in contact with my lips, I could taste no salt. I was startled by the discovery. "Father!" I fairly gasped, "the water, the water; it is fresh!" "What, Olaf?" exclaimed my father, glancing hastily around. "Surely you are mistaken. There is no land. You are going mad." "But taste it!" I cried.

And thus we made the discovery that the water was indeed fresh, absolutely so, without the least briny taste or even the suspicion of a salty flavor.

We forthwith filled our two remaining water-casks, and my father declared it was a heavenly dispensation of mercy from the gods Odin and Thor.

We were almost beside ourselves with joy, but hunger bade us end our enforced fast. Now that we had found fresh water in the open sea, what might we not expect in this strange latitude where ship had never before sailed and the splash of an oar had never been heard? (11)

(11 In vol. I, page 196, Nansen writes: "It is a peculiar phenomenon,—this dead water. We had at present a better opportunity of studying it than we desired. It occurs where a surface layer of fresh water rests upon the salt water of the sea, and this fresh water is carried along with the ship gliding on the heavier sea beneath it as if on a fixed foundation. The difference between the two strata was in this case so great that while we had drinking water on the surface, the water we got from the bottom cock of the engine-room was far too salt to be used for the boiler.")

We had scarcely appeased our hunger when a breeze began filling the idle sails, and, glancing at the compass, we found the northern point pressing hard against the glass.

In response to my surprise, my father said, "I have heard of this before; it is what they call the dipping of the needle."

We loosened the compass and turned it at right angles with the surface of the sea before its point would free itself from the glass and point according to unmolested attraction. It shifted uneasily, and seemed as unsteady as a drunken man, but finally pointed a course.

Before this we thought the wind was carrying us north by northwest, but, with the needle free, we discovered, if it could be relied upon, that we were sailing slightly north by northeast. Our course, however, was ever tending northward. (12)

(12 In volume II, pages 18 and 19, Nansen writes about the inclination of the needle. Speaking of Johnson, his aide: "One day—it was November 24—he came in to supper a little after six o'clock, quite alarmed, and said: 'There has just been a singular

inclination of the needle in twenty-four degrees. And remarkably enough, its northern extremity pointed to the east.'"

We again find in Peary's first voyage—page 67,—the following: "It had been observed that from the moment they had entered Lancaster Sound, the motion of the compass needle was very sluggish, and both this and its deviation increased as they progressed to the westward, and continued to do so in descending this inlet. Having reached latitude 73 degrees, they witnessed for the first time the curious phenomenon of the directive power of the needle becoming so weak as to be completely overcome by the attraction of the ship, so that the needle might now be said to point to the north pole of the ship.")

The sea was serenely smooth, with hardly a choppy wave, and the wind brisk and exhilarating. The sun's rays, while striking us aslant, furnished tranquil warmth. And thus time wore on day after day, and we found from the record in our logbook, we had been sailing eleven days since the storm in the open sea.

By strictest economy, our food was holding out fairly well, but beginning to run low. In the meantime, one of our casks of water had been exhausted, and my father said: "We will fill it again." But, to our dismay, we found the water was now as salt as in the region of the Lofoden Islands off the coast of Norway. This necessitated our being extremely careful of the remaining cask.

I found myself wanting to sleep much of the time; whether it was the effect of the exciting experience of sailing in unknown waters, or the relaxation from the awful excitement incident to our adventure in a storm at sea, or due to want of food, I could not say.

I frequently lay down on the bunker of our little sloop, and looked far up into the blue dome of the sky; and, notwithstanding the sun was shining far away in the east, I always saw a single star overhead. For several days, when I looked for this star, it was always there directly above us.

It was now, according to our reckoning, about the first of August. The sun was high in the heavens, and was so bright that I could no longer see the one lone star that attracted my attention a few days earlier.

One day about this time, my father startled me by calling my attention to a novel sight far in front of us, almost at the horizon. "It is a mock sun," exclaimed my father. "I have read of them; it is called a reflection or mirage. It will soon pass away."

But this dull-red, false sun, as we supposed it to be, did not pass away for several hours; and while we were unconscious of its emitting any rays of light, still there was no time thereafter when we could not sweep the horizon in front and locate the illumination of the so-called false sun, during a period of at least twelve hours out of every twenty-four.

Clouds and mists would at times almost, but never entirely, hide its location. Gradually it seemed to climb higher in the horizon of the uncertain purply sky as we advanced.

It could hardly be said to resemble the sun, except in its circular shape, and when not obscured by clouds or the ocean mists, it had a hazy-red, bronzed appearance, which would change to a white light like a luminous cloud, as if reflecting some greater light beyond.

We finally agreed in our discussion of this smoky furnace-colored sun, that, whatever the cause of the phenomenon, it was not a reflection of our sun, but a planet of some sort—a reality. (13)

(13 Nansen, on page 394, says: "To-day another noteworthy thing happened, which was that about mid-day we saw the sun, or to be more correct, an image of the sun, for it was only a mirage. A peculiar impression was produced by the sight of that glowing fire lit just above the outermost edge of the ice. According to the enthusiastic descriptions given by many Arctic travelers of the first appearance of this god of life after the long winter night, the impression ought to be one of jubilant excitement; but it was not so in my case. We had not expected to see it for some days yet, so that my feeling was rather one of pain, of disappointment that we must have drifted farther south than we thought. So it was with pleasure I soon discovered that it could not be the sun itself. The mirage was at first a flattened-out, glowing red, streak of fire on the horizon; later there were two streaks, the one above the other, with a dark space between; and from the maintop I could see four, or even five, such horizontal lines directly over one another, all of equal length, as if one could only imagine a square, dull-red sun, with horizontal dark streaks across it.")

One day soon after this, I felt exceedingly drowsy, and fell into a sound sleep. But it seemed that I was almost immediately aroused by my father's vigorous shaking of me by the shoulder and saying: "Olaf, awaken; there is land in sight!"

I sprang to my feet, and oh! joy unspeakable! There, far in the distance, yet directly in our path, were lands jutting boldly into the sea. The shore-line stretched far away to the right of us, as far as the eye could see, and all along the sandy beach were waves breaking into choppy foam, receding, then going forward again, ever chanting in monotonous thunder tones the song of the deep. The banks were covered with trees and vegetation.

I cannot express my feeling of exultation at this discovery. My father stood motionless, with his hand on the tiller, looking straight ahead, pouring out his heart in thankful prayer and thanksgiving to the gods Odin and Thor.

In the meantime, a net which we found in the stowage had been cast, and we caught a few fish that materially added to our dwindling stock of provisions.

The compass, which we had fastened back in its place, in fear of another storm, was still pointing due north, and moving on its pivot, just as it had at Stockholm. The dipping of the needle had ceased. What could this mean? Then, too, our many days of sailing had certainly carried us far past the North Pole. And yet the needle continued to point north. We were sorely perplexed, for surely our direction was now south. (14)

(14 Peary's first voyage, pages 69 and 70, says: *"On reaching Sir Byam Martin's Island, the nearest to Melville Island, the latitude of the place of observation was 75 degrees - 09' - 23", and the longitude 103 degrees - 44' - 37"; the dip of the magnetic needle 88 degrees - 25' - 56" west in the longitude of 91 degrees - 48', where the last observations on the shore had been made, to 165 degrees - 50' - 09", east, at their present station, so that we had,"* says Peary, *"in sailing*

over the space included between these two meridians, crossed immediately northward of the magnetic pole, and had undoubtedly passed over one of those spots upon the globe where the needle would have been found to vary 180 degrees, or in other words, where the North Pole would have pointed to the south.")

We sailed for three days along the shoreline, then came to the mouth of a fjord or river of immense size. It seemed more like a great bay, and into this we turned our fishing-craft, the direction being slightly northeast of south. By the assistance of a fretful wind that came to our aid about twelve hours out of every twenty-four, we continued to make our way inland, into what afterward proved to be a mighty river, and which we learned was called by the inhabitants Hiddekel.

We continued our journey for ten days thereafter, and found we had fortunately attained a distance inland where ocean tides no longer affected the water, which had become fresh.

The discovery came none too soon, for our remaining cask of water was well-nigh exhausted. We lost no time in replenishing our casks, and continued to sail farther up the river when the wind was favorable.

Along the banks great forests miles in extent could be seen stretching away on the shore-line. The trees were of enormous size. We landed after anchoring near a sandy beach, and waded ashore, and were rewarded by finding a quantity of nuts that were very palatable and satisfying to hunger, and a welcome change from the monotony of our stock of provisions.

It was about the first of September, over five months, we calculated, since our leave-taking from Stockholm. Suddenly we were frightened almost out of our wits by hearing in the far distance the singing of people. Very soon thereafter we discovered a huge ship gliding down the river directly toward us. Those aboard were singing in one mighty chorus that, echoing from bank to bank, sounded like a thousand voices, filling the whole universe with quivering melody. The accompaniment was played on stringed instruments not unlike our harps.

It was a larger ship than any we had ever seen, and was differently constructed. (15)

(15 Asiatic Mythology,—page 240, "Paradise found"—from translation by Sayce, in a book called "Records of the Past," we were told of a "dwelling" which "the gods created for" the first human beings,—a dwelling in which they "became great" and "increased in numbers," and the location of which is described in words exactly corresponding to those of Iranian, Indian, Chinese, Eddaic and Aztecan literature; namely, "in the center of the Earth."—Warren.)

At this particular time our sloop was becalmed, and not far from the shore. The bank of the river, covered with mammoth trees, rose up several hundred feet in beautiful fashion. We seemed to be on the edge of some primeval forest that doubtless stretched far inland.

The immense craft paused, and almost immediately a boat was lowered and six men of gigantic stature rowed to our little fishing-sloop. They spoke to us in a strange language. We knew from their manner, however, that they were not unfriendly. They talked a great deal among themselves, and one of them laughed immoderately, as though in finding us a queer discovery had been made. One of them spied our compass, and it seemed to interest them more than any other part of our sloop.

Finally, the leader motioned as if to ask whether we were willing to leave our craft to go on board their ship. "What say you, my son?" asked my father. "They cannot do any more than kill us."

"They seem to be kindly disposed," I replied, "although what terrible giants! They must be the select six of the kingdom's crack regiment. Just look at their great size."

"We may as well go willingly as be taken by force," said my father, smiling, "for they are certainly able to capture us." Thereupon he made known, by signs, that we were ready to accompany them.

Within a few minutes we were on board the ship, and half an hour later our little fishing-craft had been lifted bodily out of the water by a strange sort of hook and tackle, and set on board as a curiosity.

There were several hundred people on board this, to us, mammoth ship, which we discovered was called "The Naz," meaning, as we afterward learned, "Pleasure," or to give a more proper interpretation, "Pleasure Excursion" ship.

If my father and I were curiously observed by the ship's occupants, this strange race of giants offered us an equal amount of wonderment.

There was not a single man aboard who would not have measured fully twelve feet in height. They all wore full beards, not particularly long, but seemingly short-cropped. They had mild and beautiful faces, exceedingly fair, with ruddy complexions. The hair and beard of some were black, others sandy, and still others yellow. The captain, as we designated the dignitary in command of the great vessel, was fully a head taller than any of his companions. The women averaged from ten to eleven feet in height. Their features were especially regular and refined, while their complexion was of a most delicate tint heightened by a healthful glow. (16)

(16 "According to all procurable data, that spot at the era of man's appearance upon the stage was in the now lost 'Miocene continent,' which then surrounded the Arctic Pole. That in that true, original Eden some of the early generations of men attained to a stature and longevity unequaled in any countries known to postdiluvian history is by no means scientifically incredible."—Wm. F. Warren, "Paradise Found," p. 284.)

Both men and women seemed to possess that particular ease of manner which we deem a sign of good breeding, and, notwithstanding their huge statures, there was nothing about them suggesting awkwardness. As I was a lad in only my nineteenth year, I was doubtless looked upon as a true Tom Thumb. My father's six feet three did not lift the top of his head above the waist line of these people.

Each one seemed to vie with the others in extending courtesies and showing kindness to us, but all laughed heartily, I remember, when they had to improvise chairs for my father and myself to sit at table. They were richly attired in a costume peculiar to themselves, and very attractive. The men were clothed in handsomely embroidered tunics of silk and satin and belted at the waist. They wore knee-breeches and stockings of a fine texture, while their feet were encased in sandals adorned with gold buckles. We early discovered that gold was one of the most common metals known, and that it was used extensively in decoration.

Strange as it may seem, neither my father nor myself felt the least bit of solicitude for our safety. "We have come into our own," my father said to me. "This is the fulfillment of the tradition told me by my father and my father's father, and still back for many generations of our race. This is, assuredly, the land beyond the North Wind."

We seemed to make such an impression on the party that we were given specially into the charge of one of the men, Jules Galdea, and his wife, for the purpose of being educated in their language; and we, on our part, were just as eager to learn as they were to instruct.

At the captain's command, the vessel was swung cleverly about, and began retracing its course up the river. The machinery, while noiseless, was very powerful.

The banks and trees on either side seemed to rush by. The ship's speed, at times, surpassed that of any railroad train on which I have ever ridden, even here in America. It was wonderful.

In the meantime we had lost sight of the sun's rays, but we found a radiance "within" emanating from the dull-red sun which had already attracted our attention, now giving out a white light seemingly from a cloud-bank far away in front of us. It dispensed a greater light, I should say, than two full moons on the clearest night.

In twelve hours this cloud of whiteness would pass out of sight as if eclipsed, and the twelve hours following corresponded with our night. We early learned that these strange people were worshipers of this great cloud of night. It was "The Smoky God" of the "Inner World."

The ship was equipped with a mode of illumination which I now presume was electricity, but neither my father nor myself were sufficiently skilled in mechanics to understand whence came the power to operate the ship, or to maintain the soft beautiful lights that answered the same purpose of our present methods of lighting the streets of our cities, our houses and places of business.

It must be remembered, the time of which I write was the autumn of 1829, and we of the "outside" surface of the Earth knew nothing then, so to speak, of electricity.

The electrically surcharged condition of the air was a constant vitalizer. I never felt better in my life than during the two years my father and I sojourned on the inside of the Earth.

To resume my narrative of events; The ship on which we were sailing came to a stop two days after we had been taken on board. My father said as nearly as he could judge, we were directly under Stockholm or London. The city we had reached was called "Jehu," signifying a seaport town. The houses were large and beautifully constructed, and quite uniform in appearance, yet without sameness. The principal occupation of the people appeared to be agriculture; the hillsides were covered with vineyards, while the valleys were devoted to the growing of grain.

I never saw such a display of gold. It was everywhere. The door-casings were inlaid and the tables were veneered with sheetings of gold. Domes of the public buildings were of gold. It was used most generously in the finishings of the great temples of music.

Vegetation grew in lavish exuberance, and fruit of all kinds possessed the most delicate flavor. Clusters of grapes four and five feet in length, each grape as large as an orange, and apples larger than a man's head typified the wonderful growth of all things on the "inside" of the Earth.

The great redwood trees of California would be considered mere underbrush compared with the giant forest trees extending for miles and miles in all directions. In many directions along the foothills of the mountains vast herds of cattle were seen during the last day of our travel on the river.

We heard much of a city called "Eden," but were kept at "Jehu" for an entire year. By the end of that time we had learned to speak fairly well the language of this strange race of people. Our instructors, Jules Galdea and his wife, exhibited a patience that was truly commendable.

One day an envoy from the Ruler at "Eden" came to see us, and for two whole days my father and myself were put through a series of surprising questions. They wished to know from whence we came, what sort of people dwelt "without," what God we worshiped, our religious beliefs, the mode of living in our strange land, and a thousand other things.

The compass which we had brought with us attracted especial attention. My father and I commented between ourselves on the fact that the compass still pointed north, although we now knew that we had sailed over the curve or edge of the Earth's aperture, and were far along southward on the "inside" surface of the Earth's crust, which, according to my father's estimate and my own, is about three hundred miles in thickness from the "inside" to the "outside" surface. Relatively speaking, it is no thicker than an egg-shell, so that there is almost as much surface on the "inside" as on the "outside" of the Earth.

The great luminous cloud or ball of dull-red fire—fiery-red in the mornings and evenings, and during the day giving off a beautiful white light, "The Smoky God,"—is seemingly suspended in the center of the great vacuum "within" the Earth, and held to its place by the immutable law of gravitation, or a repellant atmospheric force, as the case may be. I refer to the known power that draws or repels with equal force in all directions.

The base of this electrical cloud or central luminary, the seat of the gods, is dark and non-transparent, save for innumerable small openings, seemingly in the bottom of the great support or altar of the Deity, upon which "The Smoky God" rests; and, the lights shining through these many openings twinkle at night in all their splendor, and seem to be stars, as natural as the stars we saw shining when in our home at Stockholm, excepting that they appear larger. "The Smoky God," therefore, with each daily revolution of the Earth, appears to come up in the east and go down in the west, the same as does our sun on the external surface. In reality, the people "within" believe that "The Smoky God" is the throne of their Jehovah, and is stationary. The effect of night and day is, therefore, produced by the Earth's daily rotation.

I have since discovered that the language of the people of the Inner World is much like the Sanskrit.

After we had given an account of ourselves to the emissaries from the central seat of government of the inner continent, and my father had, in his crude way, drawn maps, at their request, of the "outside" surface of the Earth, showing the divisions of land and water, and giving the name of each of the continents, large islands and the oceans, we were taken overland to the city of "Eden," in a conveyance different from anything we have in Europe or America. This vehicle was doubtless some electrical contrivance. It was noiseless, and ran on a single iron rail in perfect balance. The trip was made at a very high rate of speed. We were carried up hills and down dales, across valleys and again along the sides of steep mountains, without any apparent attempt having been made to level the Earth as we do for railroad tracks. The car seats were huge yet comfortable affairs, and very high above the floor of the car. On the top of each car were high geared fly wheels lying on their sides, which were so automatically adjusted that, as the speed of the car increased, the high speed of these fly wheels geometrically increased. Jules Galdea explained to us that these revolving fan-like wheels on top of the cars destroyed atmospheric pressure, or what is generally understood by the term gravitation, and with this force thus destroyed or rendered nugatory the car is as safe from falling to one side or the other from the single rail track as if it were in a vacuum; the fly wheels in their rapid revolutions destroying effectually the so-called power of gravitation, or the force of atmospheric pressure or whatever potent influence it may be that causes all unsupported things to fall downward to the Earth's surface or to the nearest point of resistance.

The surprise of my father and myself was indescribable when, amid the regal magnificence of a spacious hall, we were finally brought before the Great High Priest, ruler over all the land. He was richly robed, and much taller than those about him, and could not have been less than fourteen or fifteen feet in height. The immense room in which we were received seemed finished in solid slabs of gold thickly studded with jewels, of amazing brilliancy.

The city of "Eden" is located in what seems to be a beautiful valley, yet, in fact, it is on the loftiest mountain plateau of the Inner Continent, several thousand feet higher than any portion of the surrounding country. It is the most beautiful place I have ever beheld in all my travels. In this elevated garden all manner of fruits, vines, shrubs, trees, and flowers grow in riotous profusion.

In this garden four rivers have their source in a mighty artesian fountain. They divide and flow in four directions. This place is called by the inhabitants the "navel of the Earth," or the beginning, "the cradle of the human race." The names of the rivers are the Euphrates, the Pison, the Gihon, and the Hiddekel. (17)

(17 "And the Lord God planted a garden, and out of the ground made the Lord God to grow every tree that is pleasant to the sight and good for food."—The Book of Genesis.)

The unexpected awaited us in this palace of beauty, in the finding of our little fishing-craft. It had been brought before the High Priest in perfect shape, just as it had been taken from the waters that day when it was loaded on board the ship by the people who discovered us on the river more than a year before.

We were given an audience of over two hours with this great dignitary, who seemed kindly disposed and considerate. He showed himself eagerly interested, asking us numerous questions, and invariably regarding things about which his emissaries had failed to inquire.

At the conclusion of the interview he inquired our pleasure, asking us whether we wished to remain in his country or if we preferred to return to the "outer" world, providing it were possible to make a successful return trip, across the frozen belt barriers that encircle both the northern and southern openings of the Earth.

My father replied: "It would please me and my son to visit your country and see your people, your colleges and palaces of music and art, your great fields, your wonderful forests of timber; and after we have had this pleasurable privilege, we should like to try to return to our home on the 'outside' surface of the Earth. This son is my only child, and my good wife will be weary awaiting our return."

"I fear you can never return," replied the Chief High Priest, "because the way is a most hazardous one. However, you shall visit the different countries with Jules Galdea as your escort, and be accorded every courtesy and kindness. Whenever you are ready to attempt a return voyage, I assure you that your boat which is here on exhibition shall be put in the waters of the river Hiddekel at its mouth, and we will bid you Jehovah-speed."

Thus terminated our only interview with the High Priest or Ruler of the continent.

PART FOUR: IN THE UNDER WORLD

WE learned that the males do not marry before they are from seventy-five to one hundred years old, and that the age at which women enter wedlock is only a little less, and that both men and women frequently live to be from six to eight hundred years old, and in some instances much older. (18)

(18 Josephus says: "God prolonged the life of the patriarchs that preceded the deluge, both on account of their virtues and to give them the opportunity of perfecting the sciences of geometry and astronomy, which they had discovered; which they could not have done if they had not lived 600 years, because it is only after the lapse of 600 years that the great year is accomplished."—Flammarion, Astronomical Myths, Paris p. 26.)

During the following year we visited many villages and towns, prominent among them being the cities of Nigi, Delfi, Hectea, and my father was called upon no less than a half-dozen times to go over the maps which had been made from the rough sketches he had originally given of the divisions of land and water on the "outside" surface of the Earth.

I remember hearing my father remark that the giant race of people in the land of "The Smoky God" had almost as accurate an idea of the geography of the "outside" surface of the Earth as had the average college professor in Stockholm.

In our travels we came to a forest of gigantic trees, near the city of Delfi. Had the Bible said there were trees towering over three hundred feet in height, and more than thirty feet in diameter, growing in the Garden of Eden, the Ingersolls, the Tom Paines and Voltaires would doubtless have pronounced the statement a myth. Yet this is the description of the California sequoia gigantea; but these California giants pale into insignificance when compared with the forest Goliaths found in the "within" continent, where abound mighty trees from eight hundred to one thousand feet in height, and from one hundred to one hundred and twenty feet in diameter; countless in numbers and forming forests extending hundreds of miles back from the sea.

The people are exceedingly musical, and learned to a remarkable degree in their arts and sciences, especially geometry and astronomy. Their cities are equipped with vast palaces of music, where not infrequently as many as twenty-five thousand lusty voices of this giant race swell forth in mighty choruses of the most sublime symphonies.

The children are not supposed to attend institutions of learning before they are twenty years old. Then their school life begins and continues for thirty years, ten of which are uniformly devoted by both sexes to the study of music.

Their principal vocations are architecture, agriculture, horticulture, the raising of vast herds of cattle, and the building of conveyances peculiar to that country, for travel on land and water. By some device which I cannot explain, they hold communion with one another between the most distant parts of their country, on air currents.

All buildings are erected with special regard to strength, durability, beauty and symmetry, and with a style of architecture vastly more attractive to the eye than any I have ever observed elsewhere.

About three-fourths of the "inner" surface of the Earth is land and about one-fourth water. There are numerous rivers of tremendous size, some flowing in a northerly direction and others southerly. Some of these rivers are thirty miles in width, and it is out of these vast waterways, at the extreme northern and southern parts of the "inside" surface of the Earth, in regions where low temperatures are experienced, that fresh-water icebergs are formed. They are then pushed out to sea like huge tongues of ice, by the abnormal freshets of turbulent waters that, twice every year, sweep everything before them.

We saw innumerable specimens of bird-life no larger than those encountered in the forests of Europe or America. It is well known that during the last few years whole species of birds have quit the Earth. A writer in a recent article on this subject says: (19)

(19 "Almost every year sees the final extinction of one or more bird species. Out of fourteen varieties of birds found a century since on a single island—the West Indian island of St. Thomas—eight have now to be numbered among the missing.")

Is it not possible that these disappearing bird species quit their habitation without, and find an asylum in the "within world"?

Whether inland among the mountains, or along the seashore, we found bird life prolific. When they spread their great wings some of the birds appeared to measure thirty feet from tip to tip. They are of great variety and many colors. We were permitted to climb up on the edge of a rock and examine a nest of eggs. There were five in the nest, each of which was at least two feet in length and fifteen inches in diameter.

After we had been in the city of Hectea about a week, Professor Galdea took us to an inlet, where we saw thousands of tortoises along the sandy shore. I hesitate to state the size of these great creatures. They were from twenty-five to thirty feet in length, from fifteen to twenty feet in width and fully seven feet in height. When one of them projected its head it had the appearance of some hideous sea monster.

The strange conditions "within" are favorable not only for vast meadows of luxuriant grasses, forests of giant trees, and all manner of vegetable life, but wonderful animal life as well.

One day we saw a great herd of elephants. There must have been five hundred of these thunder-throated monsters, with their restlessly waving trunks. They were tearing huge boughs from the trees and trampling smaller growth into dust like so much hazel-brush. They would average over 100 feet in length and from 75 to 85 in height.

It seemed, as I gazed upon this wonderful herd of giant elephants, that I was again living in the public library at Stockholm, where I had spent much time studying the wonders of the Miocene age. I was filled with mute astonishment, and my father was speechless with awe. He held my arm with a protecting grip, as if fearful harm would overtake us. We were two atoms in this great forest, and, fortunately, unobserved by this vast herd of elephants as they drifted on and away, following a leader as does a herd of sheep. They browsed from growing herbage which they encountered as they traveled, and now and again shook the firmament with their deep bellowing. (20)

(20 "Moreover, there were a great number of elephants in the island: and there was provision for animals of every kind. Also whatever fragrant things there are in the Earth, whether roots or herbage, or woods, or distilling drops of flowers or fruits, grew and thrived in that land."—The Cratylus of Plato.)

There is a hazy mist that goes up from the land each evening, and it invariably rains once every twenty-four hours. This great moisture and the invigorating electrical light and warmth account perhaps for the luxuriant vegetation, while the highly charged electrical air and the evenness of climatic conditions may have much to do with the giant growth and longevity of all animal life.

In places the level valleys stretched away for many miles in every direction. "The Smoky God," in its clear white light, looked calmly down. There was an intoxication in the electrically surcharged air that fanned the cheek as softly as a vanishing whisper. Nature chanted a lullaby in the faint murmur of winds whose breath was sweet with the fragrance of bud and blossom.

After having spent considerably more than a year in visiting several of the many cities of the "within" world and a great deal of intervening country, and more than two years had passed from the time we had been picked up by the great excursion ship on the river, we decided to cast our fortunes once more upon the sea, and endeavor to regain the "outside" surface of the Earth.

We made known our wishes, and they were reluctantly but promptly followed. Our hosts gave my father, at his request, various maps showing the entire "inside" surface of the Earth, its cities, oceans, seas, rivers, gulfs and bays. They also generously offered to give us all the bags of gold nuggets—some of them as large as a goose's egg—that we were willing to attempt to take with us in our little fishing-boat.

In due time we returned to Jehu, at which place we spent one month in fixing up and overhauling our little fishing sloop. After all was in readiness, the same ship "Naz" that originally discovered us, took us on board and sailed to the mouth of the river Hiddekel.

After our giant brothers had launched our little craft for us, they were most cordially regretful at parting, and evinced much solicitude for our safety. My father swore by the Gods Odin and Thor that he would surely return again within a year or two and pay them another visit. And thus we bade them adieu. We made ready and hoisted our sail, but there was little breeze. We were becalmed within an hour after our giant friends had left us and started on their return trip.

The winds were constantly blowing south, that is, they were blowing from the northern opening of the Earth toward that which we knew to be south, but which, according to our compass's pointing finger, was directly north.

For three days we tried to sail, and to beat against the wind, but to no avail. Whereupon my father said: "My son, to return by the same route as we came in is impossible at this time of year. I wonder why we did not think of this before. We have been here almost two and a half years; therefore, this is the season when the sun is beginning to shine in at the southern opening of the Earth. The long cold night is on in the Spitzbergen country."

"What shall we do?" I inquired.

"There is only one thing we can do," my father replied, "and that is to go south." Accordingly, he turned the craft about, gave it full reef, and started by the compass north but, in fact, directly south. The wind was strong, and we seemed to have struck a current that was running with remarkable swiftness in the same direction.

In just forty days we arrived at Delfi, a city we had visited in company with our guides Jules Galdea and his wife, near the mouth of the Gihon river. Here we stopped for two days, and were most hospitably entertained by the same people who had welcomed us on our former visit. We laid in some additional provisions and again set sail, following the needle due north.

On our outward trip we came through a narrow channel which appeared to be a separating body of water between two considerable bodies of land. There was a beautiful beach to our right, and we decided to reconnoiter. Casting anchor, we waded ashore to rest up for a day before continuing the outward hazardous undertaking. We built a fire and threw on some sticks of dry driftwood. While my father was walking along the shore, I prepared a tempting repast from supplies we had provided.

There was a mild, luminous light which my father said resulted from the sun shining in from the south aperture of the Earth. That night we slept soundly, and awakened the next morning as refreshed as if we had been in our own beds at Stockholm.

After breakfast we started out on an inland tour of discovery, but had not gone far when we sighted some birds which we recognized at once as belonging to the penguin family.

They are flightless birds, but excellent swimmers and tremendous in size, with white breast, short wings, black head, and long peaked bills. They stand fully nine feet high. They looked at us with little surprise, and presently waddled, rather than walked, toward the water, and swam away in a northerly direction. (21)

(21 "The nights are never so dark at the Poles as in other regions, for the moon and stars seem to possess twice as much light and effulgence. In addition, there is a continuous light, the varied shades and play of which are amongst the strangest phenomena of nature."—Rambrosson's Astronomy.)

The events that occurred during the following hundred or more days beggar description. We were on an open and iceless sea. The month we reckoned to be November or December, and we knew the so-called South Pole was turned toward the sun. Therefore, when passing out and away from the internal electrical light of "The Smoky God" and its genial warmth, we would be met by the light and warmth of the sun, shining in through the south opening of the Earth. We were not mistaken. (22)

(22 "The fact that gives the phenomenon of the polar aurora its greatest importance is that the Earth becomes self-luminous; that, besides the light which as a planet is received from the central body, it shows a capability of sustaining a luminous process proper to itself."—Humboldt.)

There were times when our little craft, driven by wind that was continuous and persistent, shot through the waters like an arrow. Indeed, had we encountered a hidden rock or obstacle, our little vessel would have been crushed into kindling-wood.

At last we were conscious that the atmosphere was growing decidedly colder, and, a few days later, icebergs were sighted far to the left. My father argued, and correctly, that the winds which filled our sails came from the warm climate "within." The time of the year was certainly most auspicious for us to make our dash for the "outside" world and attempt to scud our fishing sloop through open channels of the frozen zone which surrounds the polar regions.

We were soon amid the ice-packs, and how our little craft got through. the narrow channels and escaped being crushed I know not. The compass behaved in the same drunken and unreliable fashion in passing over the southern curve or edge of the Earth's shell as it had done on our inbound trip at the northern entrance. It gyrated, dipped and seemed like a thing possessed. (23)

(23 Captain Sabine, on page 105 in "Voyages in the Arctic Regions," says: "The geographical determination of the direction

and intensity of the magnetic forces at different points of the Earth's surface has been regarded as an object worthy of especial research. To examine in different parts of the globe, the declination, inclination and intensity of the magnetic force, and their periodical and secular variations, and mutual relations and dependencies could be duly investigated only in fixed magnetical observatories.")

One day as I was lazily looking over the sloop's side into the clear waters, my father shouted: "Breakers ahead!" Looking up, I saw through a lifting mist a white object that towered several hundred feet high, completely shutting off our advance. We lowered sail immediately, and none too soon. In a moment we found ourselves wedged between two monstrous icebergs. Each was crowding and grinding against its fellow mountain of ice. They were like two gods of war contending for supremacy. We were greatly alarmed. Indeed, we were between the lines of a battle royal; the sonorous thunder of the grinding ice was like the continued volleys of artillery. Blocks of ice larger than a house were frequently lifted up a hundred feet by the mighty force of lateral pressure; they would shudder and rock to and fro for a few seconds, then come crashing down with a deafening roar, and disappear in the foaming waters. Thus, for more than two hours, the contest of the icy giants continued.

It seemed as if the end had come. The ice pressure was terrific, and while we were not caught in the dangerous part of the jam, and were safe for the time being, yet the heaving and rending of tons of ice as it fell splashing here and there into the watery depths filled us with shaking fear.

Finally, to our great joy, the grinding of the ice ceased, and within a few hours the great mass slowly divided, and, as if an act of Providence had been performed, right before us lay an open channel. Should we venture with our little craft into this opening? If the pressure came on again, our little sloop as well as ourselves would be crushed into nothingness. We decided to take the chance, and, accordingly, hoisted our sail to a favoring breeze, and soon started out like a race-horse, running the gauntlet of this unknown narrow channel of open water.

PART FIVE: AMONG THE ICE PACKS

FOR the next forty-five days our time was employed in dodging icebergs and hunting channels; indeed, had we not been favored with a strong south wind and a small boat, I doubt if this story could have ever been given to the world.

At last, there came a morning when my father said: "My son, I think we are to see home. We are almost through the ice. See! the open water lies before us."

However, there were a few icebergs that had floated far northward into the open water still ahead of us on either side, stretching away for many miles. Directly in front of us, and by the compass, which had now righted itself, due north, there was an open sea.

"What a wonderful story we have to tell to the people of Stockholm," continued my father, while a look of pardonable elation lighted up his honest face. "And think of the gold nuggets stowed away in the hold!"

I spoke kind words of praise to my father, not alone for his fortitude and endurance, but also for his courageous daring as a discoverer, and for having made the voyage that now promised a successful end. I was grateful, too, that he had gathered the wealth of gold we were carrying home.

While congratulating ourselves on the goodly supply of provisions and water we still had on hand, and on the dangers we had escaped, we were startled by hearing a most terrific explosion, caused by the tearing apart of a huge mountain of ice. It was a deafening roar like the firing of a thousand cannon. We were sailing at the time with great speed, and happened to be near a monstrous iceberg which to all appearances was as immovable as a rockbound island. It seemed, however, that the iceberg had split and was breaking apart, whereupon the balance of the monster along which we were sailing was destroyed, and it began dipping from us. My father quickly anticipated the danger before I realized its awful possibilities. The iceberg extended down into the water many hundreds of feet, and, as it tipped over, the portion coming up out of the water caught our fishing-craft like a lever on a fulcrum, and threw it into the air as if it had been a foot-ball.

Our boat fell back on the iceberg, that by this time had changed the side next to us for the top. My father was still in the boat, having become entangled in the rigging, while I was thrown some twenty feet away.

I quickly scrambled to my feet and shouted to my father, who answered: "All is well." Just then a realization dawned upon me. Horror upon horror! The blood froze in my veins. The iceberg was still in motion, and its great weight and force in toppling over would cause it to submerge temporarily. I fully realized what a sucking maelstrom it would produce amid the worlds of water on every side. They would rush into the depression in all their fury, like white-fanged wolves eager for human prey.

In this supreme moment of mental anguish, I remember glancing at our boat, which was lying on its side, and wondering if it could possibly right itself, and if my father could escape. Was this the end of our struggles and adventures? Was this death? All these questions flashed through my mind in the fraction of a second, and a moment later I was engaged in a life and death struggle. The ponderous monolith of ice sank below the surface, and the frigid waters gurgled around me in frenzied anger. I was in a saucer, with the waters pouring in on every side. A moment more and I lost consciousness.

When I partially recovered my senses, and roused from the swoon of a half-drowned man, I found myself wet, stiff, and almost frozen, lying on the iceberg. But there was no sign of my father or of our little fishing sloop. The monster berg had recovered itself, and, with its new balance, lifted its head perhaps fifty feet above the waves. The top of this island of ice was a plateau perhaps half an acre in extent.

I loved my father well, and was grief-stricken at the awfulness of his death. I railed at fate, that I, too, had not been permitted to sleep with him in the depths of the ocean. Finally, I climbed to my feet and looked about me. The purple-domed sky above, the shoreless green ocean beneath, and only an occasional iceberg discernible! My heart sank in hopeless despair. I cautiously picked my way across the berg toward the other side, hoping that our fishing craft had righted itself.

Dared I think it possible that my father still lived? It was but a ray of hope that flamed up in my heart. But the anticipation warmed my blood in my veins and started it rushing like some rare stimulant through every fiber of my body.

I crept close to the precipitous side of the iceberg, and peered far down, hoping, still hoping. Then I made a circle of the berg, scanning every foot of the way, and thus I kept going around and around. One part of my brain was certainly becoming maniacal, while the other part, I believe, and do to this day, was perfectly rational.

I was conscious of having made the circuit a dozen times, and while one part of my intelligence knew, in all reason, there was not a vestige of hope, yet some strange fascinating aberration bewitched and compelled me still to beguile myself with expectation. The other part of my brain seemed to tell me that while there was no possibility of my father being alive, yet, if I quit making the circuitous pilgrimage, if I paused for a single moment, it would be acknowledgment of defeat, and, should I do this, I felt that I should go mad. Thus, hour after hour I walked around and around, afraid to stop and rest, yet physically powerless to continue much longer. Oh! horror of horrors! to be cast away in this wide expanse of waters without food or drink, and only a treacherous iceberg for an abiding place. My heart sank within me, and all semblance of hope was fading into black despair.

Then the hand of the Deliverer was extended, and the death-like stillness of a solitude rapidly becoming unbearable was suddenly broken by the firing of a signal-gun. I looked up in startled amazement, when, I saw, less than a half-mile away, a whaling-vessel bearing down toward me with her sail full set.

Evidently my continued activity on the iceberg had attracted their attention. On drawing near, they put out a boat, and, descending cautiously to the water's edge, I was rescued, and a little later lifted on board the whaling-ship.

I found it was a Scotch whaler, "The Arlington." She had cleared from Dundee in September, and started immediately for the Antarctic, in search of whales. The captain, Angus MacPherson, seemed kindly disposed, but in matters of discipline, as I soon learned, possessed of an iron will. When I attempted to tell him that I had come from the "inside" of the Earth, the captain and mate looked at each other, shook their heads, and insisted on my being put in a bunk under strict surveillance of the ship's physician.

I was very weak for want of food, and had not slept for many hours. However, after a few days' rest, I got up one morning and dressed myself without asking permission of the physician or anyone else, and told them that I was as sane as anyone.

The captain sent for me and again questioned me concerning where I had come from, and how I came to be alone on an iceberg in the far off Antarctic Ocean. I replied that I had just come from the "inside" of the Earth, and proceeded to tell him how my father and myself had gone in by way of Spitzbergen, and come out by way of the South Pole country, whereupon I was put in irons. I afterward heard the captain tell the mate that I was as crazy as a March hare, and that I must remain in confinement until I was rational enough to give a truthful account of myself.

Finally, after much pleading and many promises, I was released from irons. I then and there decided to invent some story that would satisfy the captain, and never again refer to my trip to the land of "The Smoky God," at least until I was safe among friends.

Within a fortnight I was permitted to go about and take my place as one of the seamen. A little later the captain asked me for an explanation. I told him that my experience had been so horrible that I was fearful of my memory, and begged him to permit me to leave the question unanswered until some time in the future. "I think you are recovering considerably," he said, "but you are not sane yet by a good deal." "Permit me to do such work as you may assign," I replied, "and if it does not compensate you sufficiently, I will pay you immediately after I reach Stockholm—to the last penny." Thus the matter rested.

On finally reaching Stockholm, as I have already related, I found that my good mother had gone to her reward more than a year before. I have also told how, later, the treachery of a relative landed me in a madhouse, where I remained for twenty-eight years—seemingly unending years—and, still later, after my release, how I returned to the life of a fisherman, following it sedulously for twenty-seven years, then how I came to America, and finally to Los Angeles, California. But all this can be of little interest to the reader. Indeed, it seems to me the climax of my wonderful travels and strange adventures was reached when the Scotch sailing-vessel took me from an iceberg on the Antarctic Ocean.

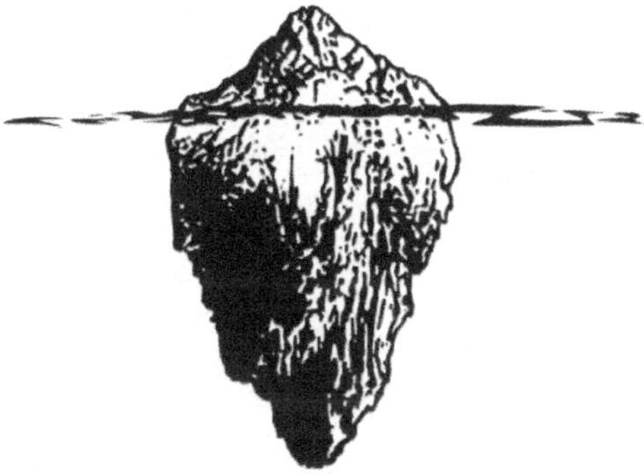

PART SIX: CONCLUSION

IN concluding this history of my adventures, I wish to state that I firmly believe science is yet in its infancy concerning the cosmology of the Earth. There is so much that is unaccounted for by the world's accepted knowledge of to-day, and will ever remain so until the land of "The Smoky God" is known and recognized by our geographers.

It is the land from whence came the great logs of cedar that have been found by explorers in open waters far over the northern edge of the Earth's crust, and also the bodies of mammoths whose bones are found in vast beds on the Siberian coast.

Northern explorers have done much. Sir John Franklin, De Haven Grinnell, Sir John Murray, Kane, Melville, Hall, Nansen, Schwatka, Greely, Peary, Ross, Gerlache, Bernacchi, Andree, Amsden, Amundson and others have all been striving to storm the frozen citadel of mystery.

I firmly believe that Andree and his two brave companions, Strindberg and Fraenckell, who sailed away in the balloon "Oreon" from the northwest coast of Spitzbergen on that Sunday afternoon of July 11, 1897, are now in the "within" world, and doubtless are being entertained, as my father and myself were entertained by the kind-hearted giant race inhabiting the inner Atlantic Continent.

Having, in my humble way, devoted years to these problems, I am well acquainted with the accepted definitions of gravity, as well as the cause of the magnetic needle's attraction, and I am prepared to say that it is my firm belief that the magnetic needle is influenced solely by electric currents which completely envelop the Earth like a garment, and that these electric currents in an endless circuit pass out of the southern end of the Earth's cylindrical opening, diffusing and spreading themselves over all the "outside" surface, and rushing madly on in their course toward the North Pole. And while these currents seemingly dash off into space at the Earth's curve or edge, yet they drop again to the "inside" surface and continue their way southward along the inside of the Earth's crust, toward the opening of the so-called South Pole. (24)

(24 "Mr. Lemstrom concluded that an electric discharge which could only be seen by means of the spectroscope was taking place on the surface of the ground all around him, and that from a distance it would appear as a faint display of Aurora, the phenomena of pale and flaming light which is some times seen on the top of the Spitzbergen Mountains."—The Arctic Manual, page 739.)

As to gravity, no one knows what it is, because it has not been determined whether it is atmospheric pressure that causes the apple to fall, or whether, 150 miles below the surface of the Earth, supposedly one-half way through the Earth's crust, there exists some powerful loadstone attraction that draws it. Therefore, whether the apple, when it leaves the limb of the tree, is drawn or impelled downward to the nearest point of resistance, is unknown to the students of physics.

Sir James Ross claimed to have discovered the magnetic pole at about seventy-four degrees latitude. This is wrong—the magnetic pole is exactly one-half the distance through the Earth's crust. Thus, if the Earth's crust is three hundred miles in thickness, which is the distance I estimate it to be, then the magnetic pole is undoubtedly one hundred and fifty miles below the surface of the Earth, it matters not where the test is made. And at this particular point one hundred and fifty miles below the surface, gravity ceases, becomes neutralized; and when we pass beyond that point on toward the "inside" surface of the Earth, a reverse attraction geometrically increases in power, until the other one hundred and fifty miles of distance is traversed, which would bring us out on the "inside" of the Earth.

Thus, if a hole were bored down through the Earth's crust at London, Paris, New York, Chicago, or Los Angeles, a distance of three hundred miles, it would connect the two surfaces. While the inertia and momentum of a weight dropped in from the "outside" surface would carry it far past the magnetic center, yet, before reaching the "inside" surface of the Earth it would gradually diminish in speed, after passing the halfway point, finally pause and immediately fall back toward the "outside" surface, and continue thus to oscillate, like the swinging of a pendulum with the power removed, until it would finally rest at the magnetic center, or at that particular point exactly one-half the distance between the "outside" surface and the "inside" surface of the Earth.

The gyration of the Earth in its daily act of whirling around in its spiral rotation—at a rate greater than one thousand miles every hour, or about seventeen miles per second—makes of it a vast electro-generating body, a huge machine, a mighty prototype of the puny-man-made dynamo, which, at best, is but a feeble imitation of nature's original.

The valleys of this inner Atlantis Continent, bordering the upper waters of the farthest north are in season covered with the most magnificent and luxuriant flowers. Not hundreds and thousands, but millions, of acres, from which the pollen or blossoms are carried far away in almost every direction by the Earth's spiral gyrations and the agitation of the wind resulting therefrom, and it is these blossoms or pollen from the vast floral meadows "within" that produce the colored snows of the Arctic regions that have so mystified the northern explorers. (25)

(25 Kane, vol. I, page 44, says: "We passed the 'crimson cliffs' of Sir John Ross in the forenoon of August 5th. The patches of red snow from which they derive their name could be seen clearly at the distance of ten miles from the coast."

La Chambre, in an account of Andree's balloon expedition, on page 144, says: "On the isle of Amsterdam the snow is tinted with red for a considerable distance, and the savants are collecting it to examine it microscopically. It presents, in fact, certain peculiarities; it is thought that it contains very small plants. Scoresby, the famous whaler, had already remarked this.")

Beyond question, this new land "within" is the home, the cradle, of the human race, and viewed from the standpoint of the discoveries made by us, must of necessity have a most important bearing on all physical, paleontological, archaeological, philological and mythological theories of antiquity.

The same idea of going back to the land of mystery—to the very beginning—to the origin of man—is found in Egyptian traditions of the earlier terrestrial regions of the gods, heroes and men, from the historical fragments of Manetho, fully verified by the historical records taken from the more recent excavations of Pompeii as well as the traditions of the North American Indians.

It is now one hour past midnight—the new year of 1908 is here, and this is the third day thereof, and having at last finished the record of my strange travels and adventures I wish given to the world, I am ready, and even longing, for the peaceful rest which I am sure will follow life's trials and vicissitudes. I am old in years, and ripe both with adventures and sorrows, yet rich with the few friends I have cemented to me in my struggles to lead a just and upright life. Like a story that is well-nigh told, my life is ebbing away. The presentiment is strong within me that I shall not live to see the rising of another sun.

Thus do I conclude my message.

Olaf Jansen

PART SEVEN: AUTHOR'S AFTERWORD

I FOUND much difficulty in deciphering and editing the manuscripts of Olaf Jansen. However, I have taken the liberty of reconstructing only a very few expressions, and in doing this have in no way changed the spirit or meaning. Otherwise, the original text has neither been added to nor taken from.

It is impossible for me to express my opinion as to the value or reliability of the wonderful statements made by Olaf Jansen. The description here given of the strange lands and people visited by him, location of cities, the names and directions of rivers, and other information herein combined, conform in every way to the rough drawings given into my custody by this ancient Norseman, which drawings together with the manuscript it is my intention at some later date to give to the Smithsonian Institution, to preserve for the benefit of those interested in the mysteries of the "Farthest North"—the frozen circle of silence. It is certain there are many things in Vedic literature, in "Josephus," the "Odyssey," the "Iliad," Terrien de Lacouperie's "Early History of Chinese Civilization," Flammarion's "Astronomical Myths," Lenormant's "Beginnings of History," Hesiod's "Theogony," Sir John de Maundeville's writings, and Sayce's "Records of the Past," that, to say the least, are strangely in harmony with the seemingly incredible text found in the yellow manuscript of the old Norseman, Olaf Jansen, and now for the first time given to the world.

THE END

Willis George Emerson

1858-1918

VARIOUS INNER OR HOLLOW EARTH DEPICTIONS

A cross-sectional drawing of the planet Earth showing the "Interior World" of *Atvatabar,* from William R. Bradshaw's 1892 science-fiction novel *The Goddess of Atvatabar.* Notices how it depicts an entry in both the North and South Poles.

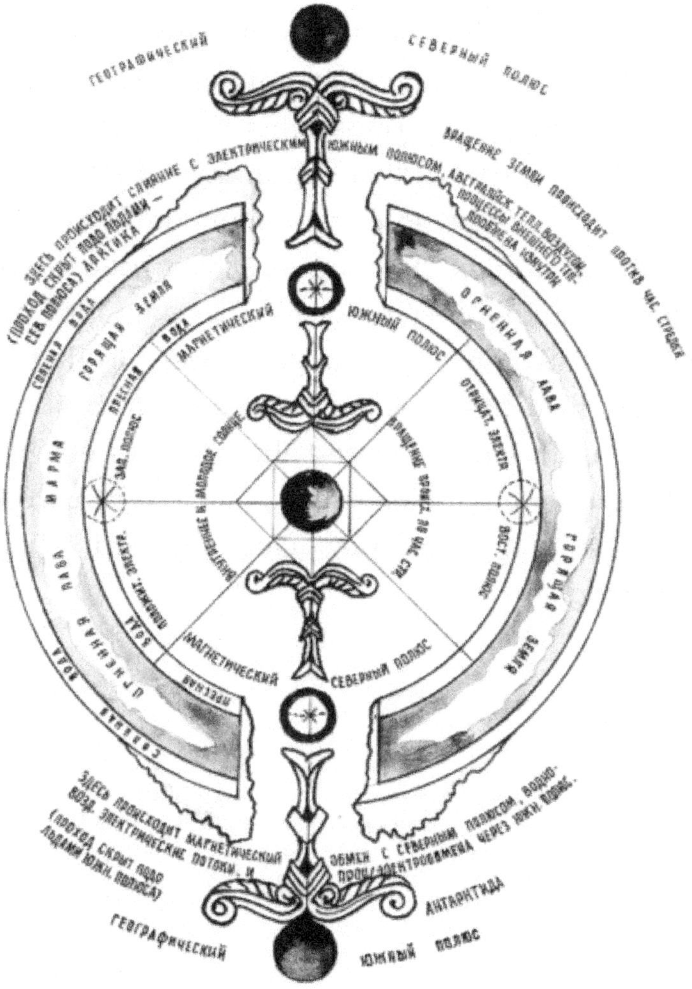

This Hollow Earth depiction is an esoteric [possibly Kabbalistic] rendition. Note at the very center the presence of a black mountain whence four diagonal lines connect. These could represent the four rivers of Eden mentioned in Genesis. Arctic exploration, with an account of Nicholas of Lynn, titled *Inventio Fortunata,* "Fortunate, or fortune-making, discovery", is a lost book, probably dating from the 14th century, containing a description of the North Pole as a magnetic island or mountain (the *Rupes Nigra,* or *Black Stone*) surrounded by a giant whirlpool, four rivers, and four continents (see detailed map on next page). Some have surmised this "Black Stone" at The North Pole could be none other than Mount Zion, the mountain where God is said to reside, spoken of abundantly in Scripture. No direct extracts from *Inventio Fortunata* have been discovered, but its influence on the Western idea of the geography of the Arctic region persisted for several centuries. Today, an elitist investment firm has adopted the name *Blackstone*. Ironically, Blackstone is the world's largest alternative asset manager, with $1 trillion in assets under management. When we follow the money, we find interesting clues. Indeed, there are *no* coincidences.

Mercator's Map: This is a famous map of the Northern Part of the World (titled *septentrio nalium terrarium de fenptio,* in Latin), by Flemish geographer and cartographer Gerardus Mercator. It is also portrayed on the cover of this book. The four rivers connecting to the center are part of Olaf Jansen's description and also of Eden in the Scriptures.

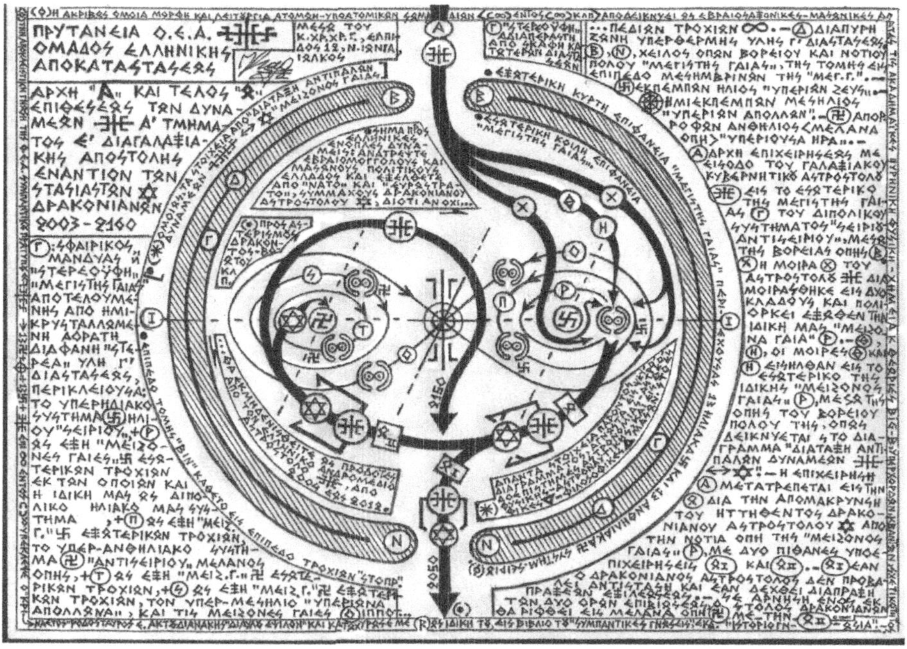

The interesting graph and Greek text of this inner Earth depiction offer some rather interesting revelations and symbolism. Note the overall Yin and Yang, male-female duality depicted. Also, and even more interesting, the six-pointed star and swastika symbols are also depicted. In modern times, we have been told they represent opposing nations, and yet, in ancient and/or occult symbology, they have more in common than one would think.

I am no Greek scholar, and unfortunately could not translate the mysterious text into English. Some commentators, however, after putting the text through an AI translator tool, said it brings to light some fascinating information.

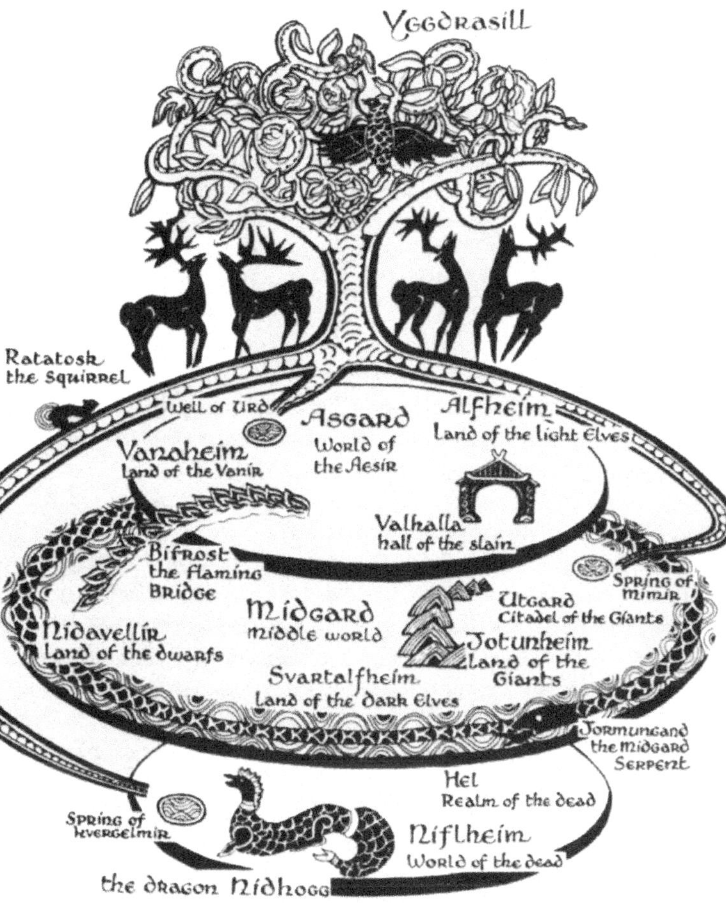

Yggdrasil (from Old Norse Yggdrasill) is an immense and central sacred tree in Norse cosmology. Around it exists all else, including the Nine Worlds of Norse mythology. According to this belief system, Midgard is Earth, and is located midway between Hel (the realm of the dead) and Asgard (the realm of the Gods—particularly Odin and Thor). This essential of Norse Mythology was part of Olaf Jansen's beliefs and Nordic heritage.

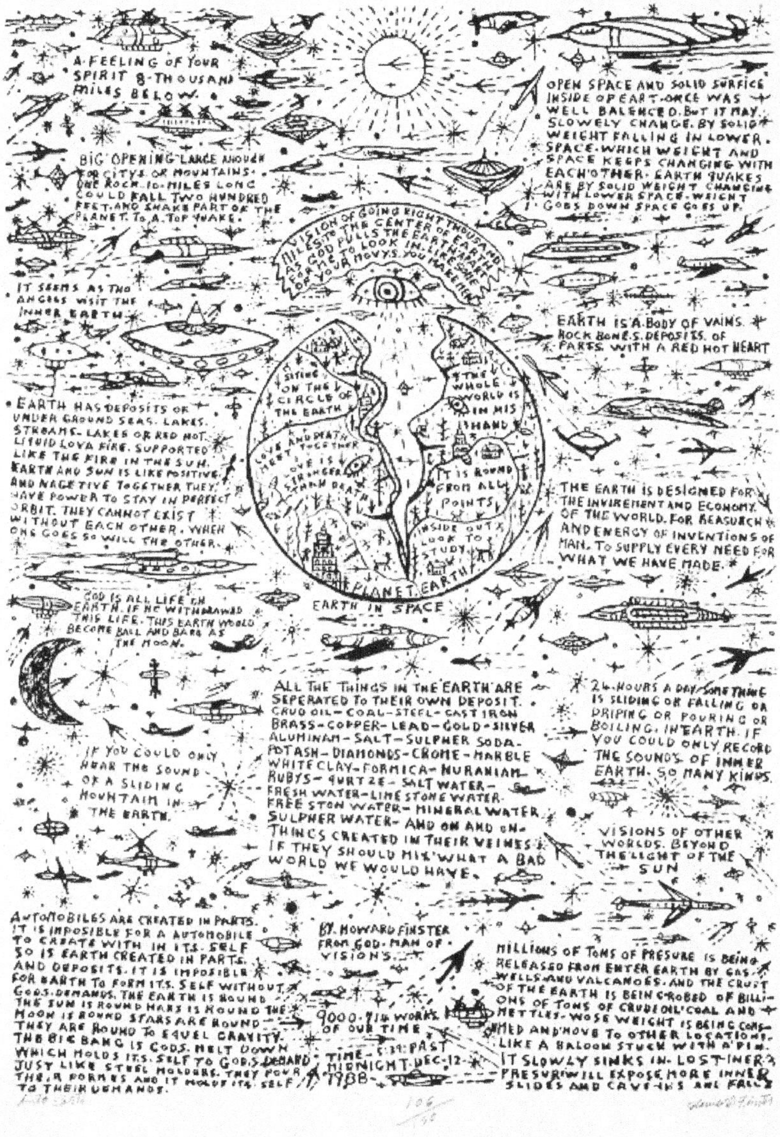

Is this an enigmatic, esoteric text and drawing of the hollow Earth? Or is this perhaps just a high school project? You be the judge.

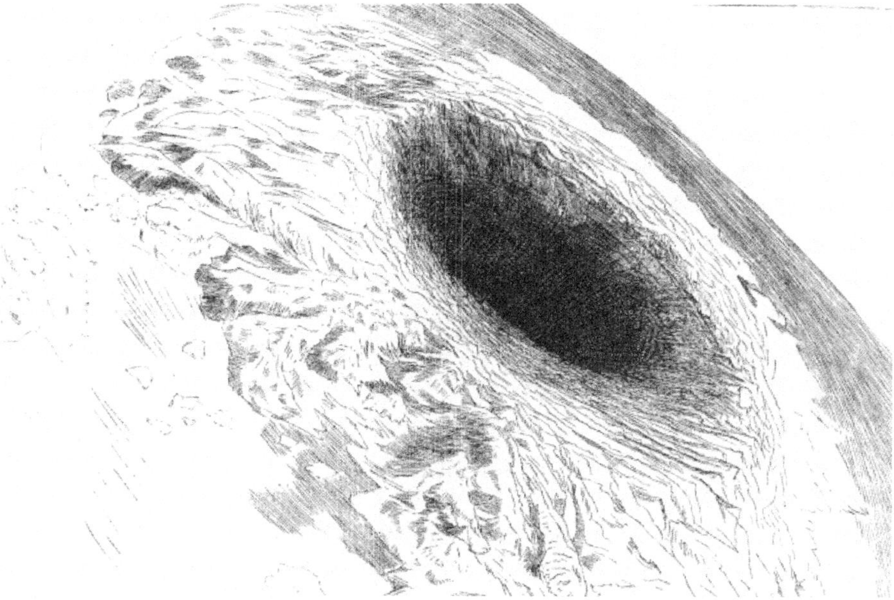

The above image is a depiction of what the North Pole and its hollow center are said to look like.

★ THE SECRET DIARY OF ADMIRAL BYRD ★

THE FLIGHT TO THE LAND BEYOND THE NORTH POLE

BY:

ADMIRAL RICHARD E. BYRD

Admiral Richard E. Byrd's Diary
(February & March 1947)

The exploration flight over the North Pole

(The Inner Earth: My Secret Diary)

I MUST write this diary in secrecy and obscurity. It concerns my Arctic flight of the nineteenth day of February in the year of Nineteen and Forty-Seven.

There comes a time when the rationality of men must fade into insignificance and one must accept the inevitability of the Truth! I am not at liberty to disclose the following documentation at this writing... perhaps it shall never see the light of public scrutiny, but I must do my duty and record here for all to read one day. In a world of greed and exploitation of certain of mankind can no longer suppress that which is truth.

FLIGHT LOG: BASE CAMP ARCTIC, 2/19/1947

0600 Hours- All preparations are complete for our flight northward and we are airborne with full fuel tanks at 0610 Hours.

0620 Hours- fuel mixture on starboard engine seems too rich, adjustment made and Pratt Whittneys are running smoothly.

0730 Hours- Radio Check with base camp. All is well and radio reception is normal.

0740 Hours- Note slight oil leak in starboard engine, oil pressure indicator seems normal, however.

0800 Hours- Slight turbulence noted from easterly direction at altitude of 2321 feet, correction to 1700 feet, no further turbulence, but tail wind increases, slight adjustment in throttle controls, aircraft performing very well now.

0815 Hours- Radio Check with base camp, situation normal.

0830 Hours- Turbulence encountered again, increase altitude to 2900 feet, smooth flight conditions again.

0910 Hours- Vast Ice and snow below, note coloration of yellowish nature, and disperse in a linear pattern. Altering course foe a better examination of this color pattern below, note reddish or purple color also. Circle this area two full turns and return to an assigned compass heading. Position check made again to base camp, and relay information concerning colorations in the Ice and snow below.

0910 Hours- Both Magnetic and Gyro compasses beginning to gyrate and wobble, we are unable to hold our heading by instrumentation. Take bearing with Sun compass , yet all seems well. The controls are seemingly slow to respond and have sluggish quality, but there is no indication of Icing!

0915 Hours- In the distance is what appears to be mountains.

0949 Hours- 29 minutes elapsed flight time from the first sighting of the mountains, it is no illusion. They are mountains and consist of a small range that I have never seen before!

0955 Hours- Altitude change to 2950 feet, encountering strong turbulence again.

1000 Hours- We are crossing over the small mountain range and still proceeding northward as best as can be ascertained. Beyond the mountain range is what appears to be a valley with a small river or stream running through the center portion. There should be no green valley below! Something is definitely wrong and abnormal here! We should be over Ice and Snow! To the portside are great forests growing on the mountain slopes. Our navigation Instruments are still spinning, the gyroscope is oscillating back and forth!

1005 Hours- I alter altitude to 1400 feet and execute a sharp left turn to better examine the valley below. It is green with either moss or a type of tight-knit grass. The Light here seems different. I cannot see the Sun anymore. We make another left turn and we spot what seems to be a large animal of some kind below us. It appears to be an elephant! NO!!! It looks more like a mammoth! This is incredible! Yet, there it is! Decrease altitude to 1000 feet and take binoculars to better examine the animal. It is confirmed—it is definitely a mammoth-like animal! Report this to base camp.

1030 Hours- Encountering more rolling green hills now. The external temperature indicator reads 74 degrees Fahrenheit! Continuing on our heading now. Navigation instruments seem normal now. I am puzzled over their actions. Attempt to contact base camp. Radio is not functioning!

1130 Hours- Countryside below is more level and normal (if I may use that word). Ahead we spot what seems to be a city!!!! This is impossible! Aircraft seems light and oddly buoyant. The controls refuse to respond!! My GOD!!! Off our port and starboard wings are a strange type of aircraft. They are closing rapidly alongside! They are disc-shaped and have a radiant quality to them. They are close enough now to see the markings on them. It is a type of Swastika!!! This is fantastic. Where are we! What has happened? I tug at the controls again. They will not respond!!!! We are caught in an invisible vice grip of some type!

1135 Hours- Our radio crackles and a voice comes through in English with what perhaps is a slight Nordic or Germanic accent! The message is: Welcome, Admiral, to our domain. We shall land you in exactly seven minutes! Relax, Admiral, you are in good hands. I note the engines of our plane have stopped running! The aircraft is under some strange control and is now turning itself. The controls are useless.

1140 Hours- Another radio message received. We begin the landing process now, and in moments the plane shudders slightly, and begins a descent as though caught in some great unseen elevator! The downward motion is negligible, and we touch down with only a slight jolt!

1145 Hours- I am making a hasty last entry in the flight log. Several men are approaching on foot toward our aircraft. They are tall with blond hair. In the distance is a large shimmering city pulsating with rainbow hues of color. I do not know what is going to happen now, but I see no signs of weapons on those approaching. I hear now a voice ordering me by name to open the cargo door. I comply.

END LOG

From this point, I write all the following events here from memory. It defies the imagination and would seem all but madness if it had not happened.

The radioman and I are taken from the aircraft and we are received in a most cordial manner. We were then boarded on a small platform-like conveyance with no wheels! It moves us toward the glowing city with great swiftness. As we approach, the city seems to be made of a crystal material. Soon we arrive at a large building that is a type I have never seen before. It appears to be right out of the design board of Frank Lloyd Wright, or perhaps more correctly, out of a Buck Rogers setting!! We are given some type of warm beverage that tasted like nothing I have ever savored before. It is delicious. After about ten minutes, two of our wondrous appearing hosts come to our quarters and announce that I am to accompany them. I have no choice but to comply. I leave my radioman behind and we walk a short distance and enter into what seems to be an elevator. We descend downward for some moments, the machine stops, and the door lifts silently upward! We then proceed down a long hallway that is lit by a rose-colored light that seems to be emanating from the very walls themselves! One of the beings motions for us to stop before a great door. Over the door is an inscription that I cannot read. The great door slides noiselessly open and I am beckoned to enter. One of my hosts speaks. Have no fear, Admiral, you are to have an audience with the Master...

I step inside and my eyes adjust to the beautiful coloration that seems to be filling the room completely. Then I begin to see my surroundings. What greeted my eyes is the most beautiful sight of my entire existence. It is in fact too beautiful and wondrous to describe. It is exquisite and delicate. I do not think there exists a human term that can describe it in any detail with justice! My thoughts are interrupted in a cordial manner by a warm rich voice of melodious quality, I bid you welcome to our domain, Admiral. I see a man with delicate features and with the etching of years upon his face. He is seated at a long table. He motions me to sit down in one of the chairs. After I am seated, he places his fingertips together and smiles. He speaks softly again and conveys the following.

"We have let you enter here because you are of noble character and well-known on the Surface World, Admiral." *Surface World*, I half-gasp under my breath! "Yes," the Master replies with a smile, 'you are in the domain of the Arianni, the

Inner World of the Earth. We shall not long delay your mission, and you will be safely escorted back to the surface and for a distance beyond. But now, Admiral, I shall tell you why you have been summoned here. Our interest rightly begins just after your race exploded the first atomic bombs over Hiroshima and Nagasaki, Japan. It was at that alarming time we sent our flying machines, the "Flugelrads ", to your surface world to investigate what your race had done. That is, of course, past history now, my dear Admiral, but I must continue on. You see, we have never interfered before in your race's wars, and barbarity, but now we must, for you have learned to tamper with a certain power that is not for man, namely, that of atomic energy. Our emissaries have already delivered messages to the powers of your world, and yet they do not heed. Now you have been chosen to be witness here that our world does exist. You see, our Culture and Science are many thousands of years beyond your race, Admiral.' I interrupted, 'But what does this have to do with me, Sir?

The Master's eyes seemed to penetrate deeply into my mind, and after studying me for a few moments he replied, "Your race has now reached the point of no return, for there are those among you who would destroy your very world rather than relinquish their power as they know it... " I nodded, and the Master continued, "In 1945 and afterward, we tried to contact your race, but our efforts were met with hostility, our Flugelrads were fired upon. Yes, even pursued with malice and animosity by your fighter planes. So, now, I say to you, my son, there is a great storm gathering in your world, a black fury that will not spend itself for many years. There will be no answer in your arms, there will be no safety in your science. It may rage on until every flower of your culture is trampled, and all human things are leveled in vast chaos. Your recent war was only a prelude of what is yet to come for your race. We here see it more clearly with each hour...do you say I am mistaken?"

'No,' I answer, 'it happened once before, the dark ages came and they lasted for more than five hundred years.'

"Yes, my son", replied the Master, "the dark ages that will come now for your race will cover the Earth like a pall, but I believe that some of your race will live through the storm, beyond that, I cannot say. We see at a great distance a new world stirring from the ruins of your race, seeking its lost and legendary treasures, and they will be here, my son, safe in our keeping. When that time arrives, we shall come forward again to help revive your culture and your race. Perhaps, by then, you will have learned the futility of war and its strife...and after that time, certain of your culture and science will be returned for your race to begin anew. You, my son, are to return to the Surface World with this message...."

With these closing words, our meeting seemed at an end. I stood for a moment as in a dream.... but, yet, I knew this was reality, and for some strange reason I bowed slightly, either out of respect or humility, I do not know which.

Suddenly, I was again aware that the two beautiful hosts who had brought me here were again at my side. 'This way, Admiral,' motioned one. I turned once more before leaving and looked back toward the Master. A gentle smile was etched on his delicate and ancient face. Farewell, my son, he spoke, then he gestured with a lovely, slender hand a motion of peace, and our meeting was truly ended.

Quickly, we walked back through the great door of the Master's chamber and once again entered into the elevator. The door slid silently downward and we were at once going upward. One of my hosts spoke again, We must now make haste, Admiral, as the Master desires to delay you no longer on your scheduled timetable and you must return with his message to your race.

I said nothing. All of this was almost beyond belief, and once again my thoughts were interrupted as we stopped. I entered the room and was again with my radioman. He had an anxious expression on his face. As I approached, I said, It is all right, Howie , it is all right. The two beings motioned us toward the awaiting conveyance, we boarded, and soon arrived back at the aircraft. The engines were idling and we boarded immediately. The whole atmosphere seemed charged now with a certain air of urgency. After the cargo door was closed the aircraft was immediately lifted by that unseen force until we reached an altitude of 2700 feet. Two of the aircraft were alongside for some distance guiding us on our return way. I must state here, the airspeed indicator registered no reading, yet we were moving along at a very rapid rate.

215 Hours- A radio message comes through. We are leaving you now, Admiral, your controls are free. Auf Wiedersehen!!!! We watched for a moment as the flugelrads disappeared into the pale blue sky.

The aircraft suddenly felt as though caught in a sharp downdraft for a moment. We quickly recovered her control. We do not speak for some time, each man has his thoughts....

ENTRY IN FLIGHT LOG CONTINUES:

220 Hours- We are again over vast areas of ice and snow, and approximately 27 minutes from base camp. We radio them, they respond. We report all conditions normal.... normal. Base camp expresses relief at our re-established contact.

300 Hours- We land smoothly at base camp. I have a mission....

END LOG ENTRIES

March 11, 1947- I have just attended a staff meeting at the Pentagon . I have stated fully my discovery and the message from the Master. All is duly recorded. The President has been advised. I am now detained for several hours (six hours, thirty-nine minutes, to be exact.) I am interviewed intently by Top Security Forces and a medical team. It was an ordeal!!!! I am placed under strict control via the national security provisions of this United States of America . I am ORDERED TO REMAIN SILENT IN REGARD TO ALL THAT I HAVE LEARNED, ON THE BEHALF OF HUMANITY. Incredible! I am reminded that I am a military man and I must obey orders.

30/12/56: FINAL ENTRY:

These last few years elapsed since 1947 have not been kind...I now make my final entry in this singular diary. In closing, I must state that I have faithfully kept this matter secret as directed all these years. It has been completely against my values of moral rights. Now, I seem to sense the long night coming on and this secret will not die with me, but as all truth shall, it will triumph and so it shall.

This can be the only hope for mankind. I have seen the truth and it has quickened my spirit and has set me free! I have done my duty toward the monstrous military-industrial complex. Now, the long night begins to approach, but there shall be no end. Just as the long night of the Arctic ends, the brilliant sunshine of Truth shall come again... and those who are of darkness shall fall in it's Light...FOR I HAVE SEEN THAT LAND BEYOND THE POLE, THAT CENTER OF THE GREAT UNKNOWN.

Admiral Richard E. Byrd

United States Navy, 24 December 1956

MORE INNER EARTH DEPICTIONS

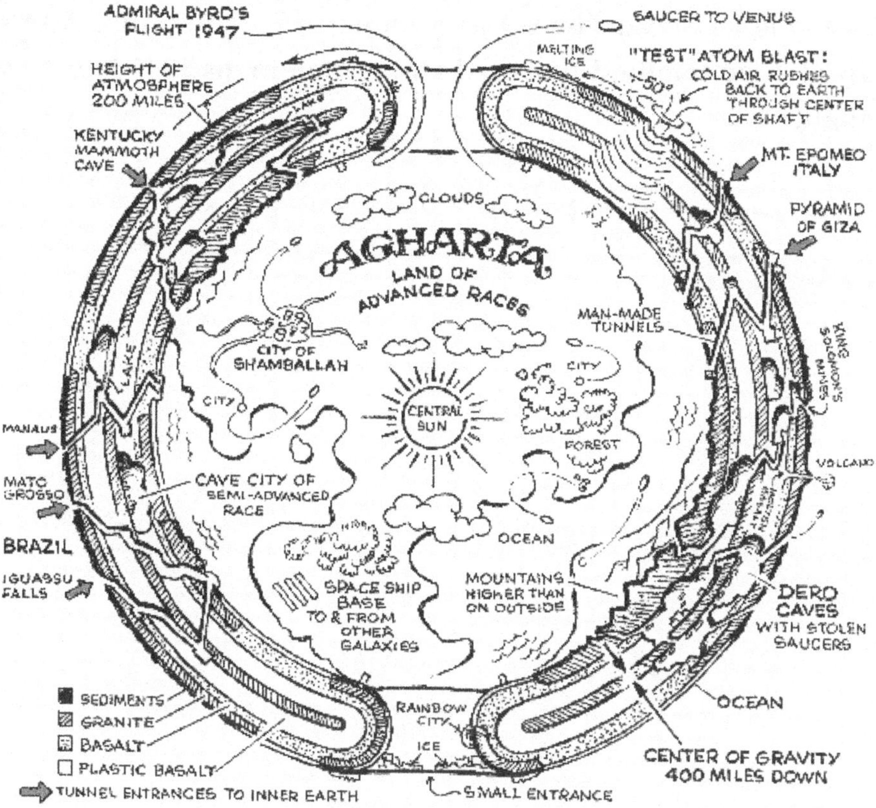

Agartha is the name given to a legendary kingdom that is said to be located on the inner surface of the Earth. It is related to the belief in a hollow/inner Earth and is a popular subject in esotericism—particularly in Eastern Mysticism. The legend of Agartha remained mostly obscure in Europe until Gérard Encausse edited and re-published a detailed 1886 account by the nineteenth-century French occultist Alexandre Saint-Yves d'Alveydre (1842–1909), *Mission de l'Inde en Europe,* in 1910. [Wikipedia]

After World War I, German occultist groups such as the Thule Society took an interest in Agartha. [Wikipedia]

Many believe Agartha is the inner earth kingdom that Admiral Byrd accidentally discovered and visited in his 1947 North Pole flight.

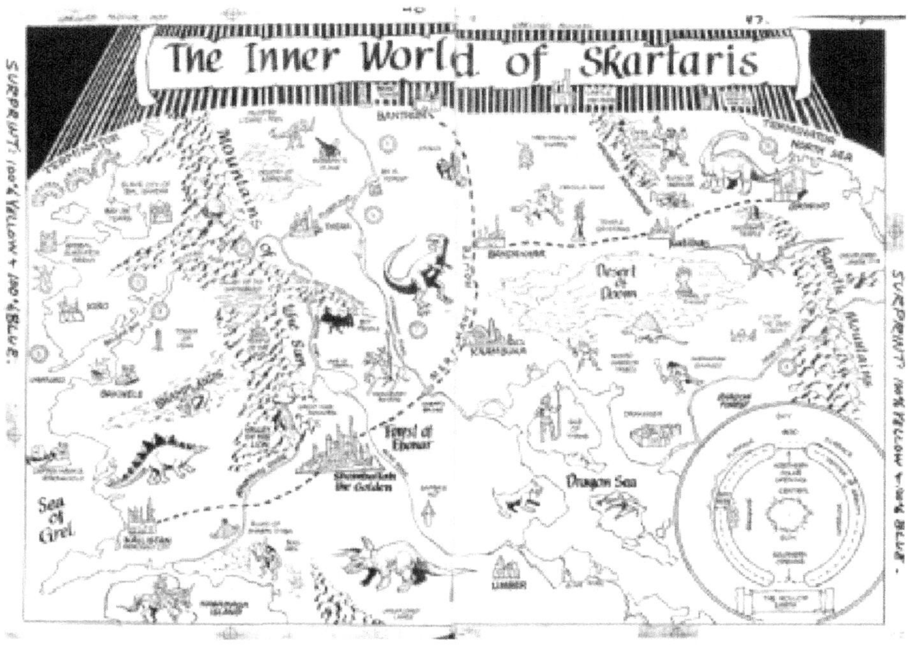

According to Wikipedia, *Skartaris* is a fictional Hollow Earth fantasy setting created by Mike Grell for the sword and sorcery comic book *The Warlord*, published by DC Comics. *Skartaris* debuted in 1st Issue Special #8 (November 1975), where the character Travis Morgan, a U.S. Air Force pilot, discovers a passage into this world through the Earth's North Pole. Subsequent to that first issue, the Warlord series tells of Morgan's adventures in Skartaris. Hmmm, I wonder where the author got such an idea?

DOCUMENTS AND PICTURES FEATURING ADMIRAL RICHARD E. BYRD

San Francisco Chronicle

LEADING NEWSPAPER of the PACIFIC COAST

FOUNDED 1865 — SAN FRANCISCO, CAL., MONDAY, MAY 10, 1926 — DAILY 5 CENTS, SUNDAY 10 CENTS

BYRD CIRCLES NORTH POLE IN PLANE

Flies 1620 Miles in 16 Hours on Nonstop Dash

Leak in Oil Line Menaces Airship At Top of World

Commander and Pilot Alternate in Handling Machine on Trip; Perfect Weather Aids in Verifying Peary's Observations

Wins Honor For America

CHURCH FRONT WRECKED IN SECOND BOMB BLAST PLOT

Parish Priests of SS. Peter and Paul Alarmed, but Escape Injury

Work of Vandals Who Planted Dynamite There Jan. 31, Police Believe

C. S. HOWARD'S SON KILLED UNDER TRUCK NEAR WILLITS

Week-End Fishing Party of Three Boys Brings Death to Its Leader

Parents Charter Train and Hurry North From San Francisco

Fishing Jaunt Proves Fatal

British Workers' Third Army Waits Orders to Strike

Walkout of General Laborers to Swell Unionist Ranks to Five Million; Negotiations Remain at Standstill

ODDITIES IN THE NEWS OF THE DAY

R E Byrd

Admiral Richard Byrd and Floyd Bennett receive a Medal of Honor from U.S. president Calvin Coolidge (center).

MYSTERIOUS FLAT EARTH MAPS

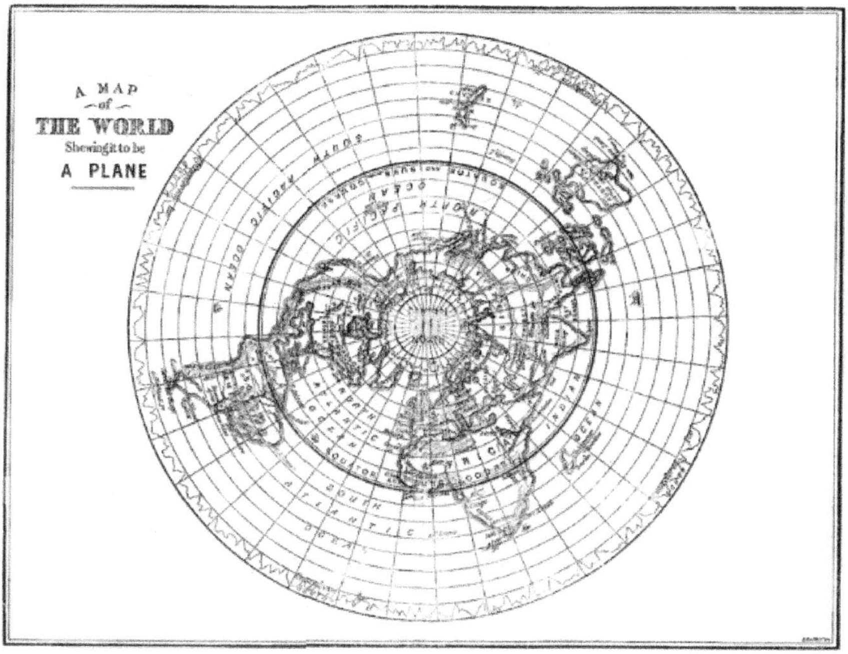

Some believe the flight of Admiral Byrd into the North Pole's interior is a more achievable feat on a Flat Earth (depicted here) than on a globe. While this is certainly debatable, Byrd's claim that *"There is left in the world today, an area as big as the United States, that's never been seen by a human being: And that's beyond the pole, on the other side of the south pole from Little America,"* is, to say the least, quite intriguing. This particular quote of the famed Arctic explorer is often used as an argument to support the circular Antarctic wall view of the Flat Earth—and the possibility of more lands beyond it.

According to the Friday, January 11, 1907 edition of the Hawaiian Gazette, this map is over 1,100 years old. It was found in a Japanese temple in the mountains of Japan. In the article, Dr. Kobayashi, a well-known Japanese physician and surgeon in Honolulu said it was sent to him by his brother, a researcher and flat-Earth adherent. It is reminiscent of the Mercator projection map (included inside this book and on the cover). But the most interesting part of this particular "flat-Earth" map is that is shows more land beyond the Antarctic border (which if the circular edge according to flat Earth beliefs). Could this be the reason the Antarctic borders are so heavily guarded by an international treaty and coalition of nations? Are only an elite few privy to such information? And, if such lands do exist, who has access to it? Who has seen them?

You can read the full article here:
https://chroniclingamerica.loc.gov/lccn/sn83025121/1907-01-11/ed-1/seq-2/

This 1587 Flat Earth Map of the World by Urbano Monte was featured in National Geographic Magazine and has baffled experts because it shows much extra land. Notice also the interesting depiction of the North Pole with what appears to be a large mountain at its center—possibly the mysterious *Rupes Nigra* (?).

ALSO AVAILABLE FROM WILD REMNANT PUBLISHING:

THE BOOK OF ENOCH

The Most Revered, Essential, and Scholarly Translation of the Book of Enoch Available for Today's Most Scrutinizing Lay Readers, Students, and Scholars

The Book of First Enoch is a pseudepigraphal Hebrew apocalyptic religious text attributed by tradition to Enoch,

Noah's great-grandfather, the seventh man from Adam. Offering a wealth of pre-diluvian and eschatological content, you'll find within its pages insights into the origins of demons and the Nephilim, an exploration of why certain angels (the Watchers) fell from heaven, a moral justification for the Genesis flood, much disclosure about our earth realm, and revelations about the Messiah's thousand-year reign.

Immerse yourself in the profound wisdom and prophecies of the First Book of Enoch with this 1917 authorized English translation by renowned Master of Ancient texts, Robert Henry (R.H.) Charles.

Featuring a thought-provoking introduction by Sebastien Richard; this unique version of the pseudepigraphal work will appeal to a wide spectrum of readers: beginners, seekers, readers of ancient texts, students, lovers of wisdom, scholars, and conspiracy researchers.

It includes these unique features:

★ **Large Print Edition:** This edition features an easy-on-the eyes 14-point type font, for added ease of reading and study.

★ **Unabridged Text:** Dive deep into the mystical world of Enoch with the complete classic English text of 1 Enoch as it was meant to be read by its ancient biblical author.

★ **Modern Chapter Numeration:** The chapter numeration was updated by replacing the translator's original Roman numerals with regular numbers. This makes for easy navigation for today's modern readers.

★ **Explanatory Chapter Subheads:** Navigate the depths of 1 Enoch with ease, thanks to carefully written explanatory chapter subheadings that provide invaluable introduction, context, and insights into the text.

Delve deep into the antediluvian world where angels, giants, and celestial realms intertwined with the human story. Unearth the timeless wisdom and hidden truths of Enoch in this unique edition. Embark on a faith journey through the ages, exploring the profound teachings and revelations of this highly esteemed extra-biblical text!

"The relevancy of The Book of Enoch at this time cannot be understated."

"Whether you're just curious, a student, or a scholar, this book will shock you."

"Enoch should be mandatory reading for all students of Bible prophecy."

AVAILABLE ON:

amazon.com

Made in United States
North Haven, CT
11 November 2026

22198992R00078